QUANTITATIVE GEOLOGY

Based on a symposium held at the 82nd Annual Meeting of The Geological Society of America, Atlantic City, New Jersey, November 10, 1969

Edited by PETER FENNER

with articles contributed by:

W. C. Krumbein, Richard B. McCammon,
John M. Dennison, Geoffrey S. Watson,
Richard A. Park, John W. Wilkinson, John C. Davis,
Floyd W. Preston, Peter J. R. Buttner, and
John C. Griffiths

THE GEOLOGICAL SOCIETY OF AMERICA

SPECIAL PAPER 146

PUBLISHED BY
THE GEOLOGICAL SOCIETY OF AMERICA, INC.
3300 PENROSE PLACE
BOULDER, COLORADO 80301

Printed in The United States of America

*The printing of this volume has been made possible
through the bequest of
Richard Alexander Fullerton Penrose, Jr.*

Foreword

In the fall of 1966, W. C. Krumbein gave a lecture at the University of Pennsylvania. From the germ of an idea provoked during an after-lecture discussion, aided and abetted by the considerable energies and cooperation provided by many people and organizations since then, a sequence of events culminating in this publication has taken place. This Quantitative Geology Sequence was in reality a curriculum package that included an annotated bibliography, two short reviews, two short courses, and the symposium that lead to this publication. Each element in the Sequence was planned to build upon previous ones and yet, given a sufficiently strong starting point, to be able to stand alone.

Most of the support for the Sequence was provided by the National Science Foundation through the Council on Education in the Geological Sciences (CEGS), a project of the American Geological Institute (AGI).

The *Bibliography of Statistical Applications in Geology,* by J. C. Howard, appeared in 1968 as CEGS Programs Publication no. 2. It was prepared under the direction of J. W. Harbaugh, with staff assistance by D. M. Delo, Jr.

Planning sessions, helped especially by W. C. Krumbein, R. B. McCammon, H. McCammon, G. V. Middleton, J. C. Snyder, G. M. Friedman, D. Merriam, W. W. Hambleton, R. Park, and officers and staff of The Geological Society of America, American Geological Institute, and National Association of Geology Teachers, led to other components in the Sequence.

In the October 1969 issue of *Journal of Geological Education* there appeared two CEGS-sponsored short reviews: "Undergraduate Instruction in Geomathematics" by R. H. Osborne, and "Computer-Oriented Laboratory Exercises for Geology and Oceanography," by W. T. Fox.

During November 1969, the Philadelphia office of Service Bureau Corporation gave a free course entitled "Introduction to Time-Sharing Computers and the PL–1 Language" to interested geologists headed for The Geological Society of America Annual Meeting in Atlantic City. This was followed by an intensive AGI/CEGS pre-GSA meeting short course with notes entitled "Models of Geologic Processes—An Introduction to Mathematical Geology," edited by P. Fenner and published by AGI. Contributions in that publication range from concepts of modelling in geology, to univariate and multivariate analytical techniques, APL matrix analysis, and computer-based graphic model production. In addition to the published contributions by W. C. Krumbein, M. E. Kauffman, R. B. McCammon, D. B. McIntyre, and H. N. Pollack, there were fine oral contributions by W. James. Some of the major calculator companies (including Hewlett-Packard, Monroe, Olivetti-Underwood, Singer-Friden, Smith-Corona-Marchant, Victor, and Wang) made several of their latest electronic models available for use of short-course participants.

The following day the symposium that forms the basis for the present publication was held. Kudos are due my co-chairman, G. V. Middleton, for his patience with my long-distance telephone calls and the many manuscripts sent to him for review. R. B. McCammon's near-to-last-minute willingness to substitute for W. C. Krumbein, when the latter had to make an emergency trip abroad, is gratefully acknowledged. Papers of both authors are included.

It should here be noted that the entire focus of the Quantitative Geology Sequence was clearly distinct from the excellent papers collected in special issues of the *Journal of Geology* and from the well-managed Kansas Geological Survey—AAPG publications and symposia. All phases of the Sequence led toward the notion of utilization of models in geology. Hence, the underlying theme of the symposium was "Models in Geology: In the Past, Present Use, and Future Directions."

Members of the CEGS/Professional Development Panel who suffered through and sanctioned the growing pains of this, the first of three planned CEGS-sponsored Sequences, included R. J. Weimer, D. R. Baker, D. S. Gorsline, V. E. Gwinn, W. R. Muehlberger, and K. Oles. Without the help, criticism, and encouragement of many others, neither this publication nor its predecessor components in the Sequence ever would have come to pass. Among the most persistent and helpful colleagues was W. Cochran.

Because some critical reviewers have preferred to remain anonymous, I wish to thank collectively those many reviewers whose hours of toil have certainly improved all the papers contained in *Special Paper 146*. This complex sequence involved author's critics, staff support, and about 75 critics and contributors. I am apologetic to anyone who feels slighted, heartened by the considerable assistance received throughout by so many, grateful to those who were able to remain on schedule, or nearly so, and pleased with the over-all results of this undertaking.

P. Fenner

Governors State University
Park Forest South, Illinois 60466

Contents

Probabilistic Models and the Quantification Process in Geology

W. C. Krumbein
Department of Geological Sciences, Northwestern University
Evanston, Illinois 60201

ABSTRACT

The introduction into geology of probabilistic models such as Markov chains for simulating stratigraphic sections, and independent-events models of drainage networks in geomorphology, calls attention to the need for re-examination of operational definitions used in conventional field and other kinds of measurement techniques. Critical examination of observed frequency distributions that arise from the measurements is especially important in terms of the physical processes that generate them. For example, some simulation models in stratigraphy give rise to geometrically distributed thicknesses of rock units. Unless the input is similarly distributed, the model does not reflect real-world events.

The quantification process involves at least five steps: (1) selection of qualities to be quantified; (2) development of operational definitions for obtaining the numbers; (3) evaluation of objectivity, accuracy, and precision of the resulting numbers; (4) examination of the frequency distributions of the numbers; and (5) critical evaluation of the usefulness of the numbers in the development and testing of geological models. Items (4) and (5) are emphasized in this paper, in terms of their bearing on the growing use of probabilistic models in geology.

INTRODUCTION

Quantification in geology is a process in which descriptive statements about qualities of geological objects are converted into numbers that express how much of the quality is possessed by the object. Once numbers are available, the variability in geological data becomes formally measurable. With proper statistical design, a within-measurements sum of squares and a within-samples sum of squares can be removed from the data, leaving a residual sum of squares that can be used to estimate the population variability. This last is an estimate of the inherent variability among the objects themselves, as, for example, in the visible differences in the size and shape of pebbles within a gravel deposit.

It is normal practice in geology to prepare histograms or cumulative curves of measurement data, to find whether the distribution is symmetrical or skewed and whether it is unimodal or not, and to relate the observed distribution to some standard density such as the normal or lognormal. It is common knowledge that different methods of measuring pebble size, say, or the shape of drainage basins, yield different values for even the same object. To what extent does the choice of operational definition influence the geological meaningfulness and physical significance of the resulting numbers? What is the influence of the operational definition on the form and estimated parameters of the resulting frequency distributions?

In my opinion neither of these questions has been adequately examined in the framework of quantitative geology. Yet their importance in geological model building may be large. In a deterministic sense, where physically meaningful systematic relations are sought between a dependent variable Y and a set of independent variables X_1, X_2, . . . , the criterion of an adequate model is that it should be able to predict Y with probability $p = 1$ when selected values of the Xs are inserted. If deviations from the predicted value are very small, the departure from exact prediction can be attributed to some content of unidentified "noise," but when the deviations are large, the researcher may conclude that his model is inappropriate. How often does this occur simply because of poor operational definitions?

When probabilistic models (stochastic process models) are invoked, we seek for underlying probabilistic mechanisms in the process, rather than focusing entirely on deterministic controls for Y. Probabilistic mechanisms, like their deterministic counterparts, behave in a rigorous mathematical manner, but their introduction into a natural process significantly changes the predictive framework in which the process is viewed. The variability in the data now tends to conform to certain laws of probability, but even here poor operational definitions may obscure the nature of the built-in randomness of the model.

PREDICTION IN GEOLOGY

Several variants of the general linear model (Krumbein and Graybill, 1965, chapters 12–14) are widely used as statistical predictive devices in geology. These include trend surface analysis, multiple regression, factor analysis, and discriminant functions, which are used for detecting and evaluating systematic relations among variables. Relative effectiveness is gaged by the percentage of the total sum of squares of Y accounted for by the technique. This is normally in the range of 75 to 95 percent, so that the predictions include varying degrees of uncertainty. In these uses of the general linear model the exact physical meaning of the coefficients may not be discernible, but for purely predictive purposes this is not necessary. Despite the element of uncertainty and the lack of substantive significance, this approach has proved its value in the search for oil, gas, and ores, especially by trend-surface analysis.

Basically the geologist is interested in understanding why geological processes operate as they do, and for this purpose the deterministic approach has proved fruitful. As in classical mathematical physics, one normally starts with a differential equation which is integrated in terms of known or assumed boundary conditions. The constants and parameters in the equation are derived from theory or found by experiment. The final equation is a deterministic model in that the value of Y is completely determined in time or space by its functional relation with the Xs. Stokes' Law and Darcy's Law are well-known examples of such models.

The computer has greatly extended the geologist's capability for developing deterministic models that involve a multiplicity of variables, but this becomes a self-limiting process because interactions may develop between the independent variables, partial dependencies rather than full dependencies may manifest themselves between initially dependent and independent variables, and feedback loops may become prominent in the system under study. The consequence of these complications is that the one-to-one relation between specific controls and responses (causes and effects) becomes blurred, so that more than one control may lead to the same response, or more than one response may result from a given control.

What happens here is that conditional probabilities enter the picture; that is, the occurrence of a particular response may depend upon whether or not some other response has already occurred. When this happens, predictions of Y from a set of Xs move out of the deterministic realm of prediction with probability $p = 1$ to a realm in which $0 < p < 1$.

Watson (1969) discusses the philosophy of prediction in geology by pointing out that when $p = 1$ changes to $0 < p < 1$, prediction changes from a deterministic to a probabilistic framework. In this changed framework predictions apply to a group of particulars instead of one particular. This difference in the predictive framework expresses a basic difference between deterministic and probabilistic models. Instead of being able to say, "Given certain values of the Xs, Y will have a specific value with probability $p = 1$," we are constrained to say that the value of a particular Y now depends on an underlying set of probabilities associated with a class or group of possible responses.

PROBABILISTIC MODELS

A single six-sided die can be used to illustrate a completely probabilistic mechanism, the independent-events model. After a given toss, any number of spots from 1 to 6 may occur on the top face. No doubt the energy of the toss, the nature of the gambling surface, the directions of roll of the die, and other factors combine in some way to control the specific response. It is doubtful whether a differential equation can be written for this die experiment inasmuch as the individual microcauses and microeffects cannot be unequivocally related to each other. Within the complex interplay of controls and responses, we do know that there are only six possible outcomes, with probability $p = 1$ that some one of the six sides must lie face up. This is a group prediction based on the total sample space for the die; within the group the most reasonable prediction we can make is that the probability is one-sixth that any particular side will lie face up. Experience has also amply shown that the outcome of any given toss is completely independent of the previous outcomes.

We can carry this probabilistic reasoning into geology for those situations in which microeffects and microcauses are intermingled by interactions and by feedback loops, and especially where a single response can arise from any of several controls. One such situation is the process of stream channel bifurcation along the growing edges of a drainage network. Bifurcation can arise because a boulder is undermined in the bank, a tree falls over, or a cowpath is washed out in a storm. A decade or less after the event, it may not be possible to identify the microcause, but this does not mean that the processes of nature have been violated. The channel still cuts uphill and the water still flows downhill. What has happened is simply that we are now forced to say that the bifurcation could have arisen in any of several ways. If we could identify the set of possible microcauses, we could then collect data in newly developing drainage areas of the relative numbers of times that one event or another takes place. In such studies we would be identifying the sample space for bifurcations and associating relative probabilities with each.

An outstanding example of an independent-events model is Shreve's (1966, 1967) introduction into geomorphology of the concept of topologically random stream channel networks. Shreve postulates that all topologically distinct channel networks having a given number of sources (Strahler first-order streams) are equally likely to occur in the absence of geological controls such as folded rocks. An important part of Shreve's contribution is the introduction of the concept of basin magnitude, which differs from basin order in that the magnitude of a basin or of a network directly expresses the number of sources (first-order streams) in the unit of interest. Inasmuch as the sample space of possible topological arrangements for basins of any magnitude can be derived, Shreve laid the foundation for complete re-examination of drainage patterns, relations between stream length and basin area (Shreve, 1969), and so on. In fact, under Shreve's postulate, Horton's Law of stream numbers is simply the outcome of an independent-events mechanism. Other workers (Smart, 1968, 1969; James and Krumbein, 1969) have built on Shreve's foundation, and although some departures from complete randomness have been noted, the soundness of the probabilistic approach is amply supported.

Normally probabilistic mechanisms in geology do contain partial dependencies, as when an immediately preceding event controls the probability of occurrence of the next event. Such partial dependencies may be part of the initial process, or they may arise in independent-events mechanisms as a growing system matures through time. The probabilistic mechanism commonly used to study partial dependencies is the Markov chain, though other kinds of models may also be used. Markov models have been most widely used in stratigraphic analysis. In cyclical deposits, for example, it is possible to discern a pattern of lithologic succession that can be expressed as an "ideal cycle" for the sequence, but predictions of the kind of lithology that immediately follows a given occurrence can only be made in terms of an underlying set of probabilities. These are arranged as a matrix of transition probabilities, p_{ij}, which express the likelihood that the system will be in state j at time t_{n+1}, given that it was in state i at time t_n. Here the states are defined as the lithologic components in the stratigraphic sequence.

It is possible statistically to test the hypothesis of an independent-events model against the alternative of a first-order Markov chain (Anderson and Goodman, 1957; *see also* Krumbein, 1967). If the hypothesis is rejected, the process presumably is Markovian, but an additional requirement needs to be satisfied. This is that the input data on the thicknesses of the lithologic units must be geometrically distributed in a discrete sense (Krumbein and Dacey, 1969), equivalent to the exponential distribution in a continuous sense. This requirement is in apparent conflict with a large body of observational data, which suggests that the thicknesses are lognormally distributed. If lognormalcy applies, the Markov model can be recast as an embedded Markov chain, as used by Potter and Blakely (1968), that belongs to the general class of semi-Markov models which are independent of the kinds of thickness distributions shown by the input data. On the other hand, if operational definitions for measuring thicknesses are insensitive to very thin units, the observed thicknesses would appear to be lognormal even though the "true" population distribution is exponential. This kind of situation shows how an operational definition could confuse or obscure an otherwise simple picture.

The use of Markov models definitely requires much closer scrutiny of observed frequency distributions than is commonly accorded them. The same holds for other probabilistic models; independent-events processes, for example, also give rise to geometric distributions in a discrete sense, as is brought out later. Emphasis thus shifts from descriptive aspects of observed frequency distributions to consideration of the particular kinds of population densities that are generated by the probabilistic mechanisms postulated in the models. In short, new concepts and radical changes in the predictive framework associated with probabilistic models heighten the need for re-examining some aspects of the quantification process itself.

QUANTIFICATION PROCESS

Numerous workers have concerned themselves with problems of measurement in geology. Some elements of the quantification process are discussed in Krumbein and Graybill (1965, chapter 3) and by Griffiths (1967, chapter 4). Griffiths gives a tabulation of the stages in quantification (p. 45), but the five aspects of quantification given in the abstract to this paper are adequate for present purposes.

The selection of qualities to be quantified is entirely a geological decision based on one's judgment regarding the kinds of data that bear on the study undertaken. The development or adaptation of operational definitions for obtaining the numbers normally arises from an attempt to ensure objectivity in the measurement process, combined with some acceptable level of accuracy and precision to avoid undue measurement error.

Frequency Distributions

Inasmuch as many measurements in geology are based on samples or test specimens of rocks, fossils, and the like, the resulting numerical data can be displayed as frequency distributions of one sort or another. Sample statistics of the observed data are influenced by the accuracy of the measurements (which in part controls the observed mean value), by precision (which affects the observed variance), by the inherent magnitude and variability of the objects and qualities being measured (which affects the mean, variance, and asymmetry), and by the specific formulation of the operational definition, which plays a strong part in geological meaningfulness of the resulting numbers, as well as in the form of the resulting frequency distributions.

This last point is brought out in Figure 1, which compares two operational

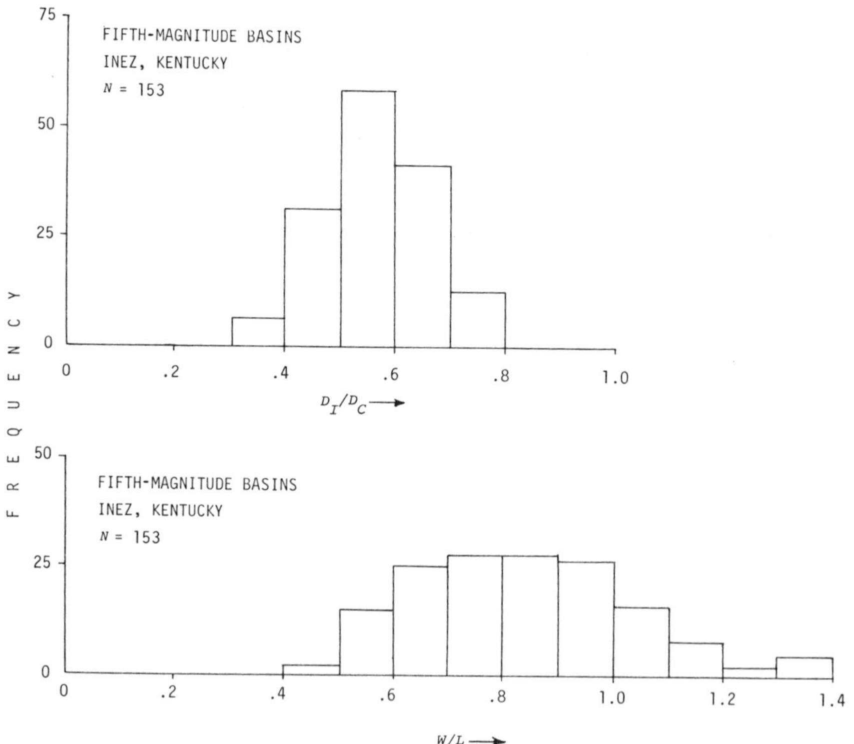

Figure 1. Comparison of frequency distributions arising from two operational definitions for drainage basin shape. See text for details.

definitions for drainage basin shape. The upper histogram is based on the ratio D_I/D_C of the diameter of the inscribed circle to that of the circumscribed circle for each basin, and the lower histogram is based simply on the ratio W/L of the width to the length of each basin. The operational definitions for these measurements are illustrated in Krumbein and Graybill (1965, p. 41–43).

The histograms are based on the same 153 fifth-magnitude basins in the Inez quadrangle (eastern Kentucky), and though the intent here is simply to use them as illustrations, some implications of the definitions may be mentioned. The upper histogram is more compact than the lower, partly because the upper limit of D_I/D_C is 1.0, whereas W/L may exceed this value for basins that are broader than long. The lower histogram is right-skewed, which is characteristic of other basin properties (such as area and total stream length), and the upper histogram is more symmetrical, almost suggesting a normal distribution.

There are strong advantages in having a normal frequency distribution (or one that can be approximately normalized by a simple transformation) for conventional statistical analysis of geological data. If adequate material is available, and if the cost of a measurement is not large, normalcy can be fairly well assured if the statistical models are entered with sample means rather than with individual measurements. This follows from the fact that the distribution of sample means based on independent drawings of n individuals each from almost any kind of distribution will tend toward normalcy as n increases.

This process, of considerable value in some kinds of statistical analysis, can be fatal when probabilistic (that is, stochastic process) models are used. For these models it is imperative that the frequency distributions of the variables involved be consistent with the requirements of the models. This may be effectively demonstrated with the six-sided die. Figure 2, left, shows the discrete population density for the die-rolling process, with equal likelihood that any of the six faces will lie face up on any given toss. The density is rectangular, with relative proba-

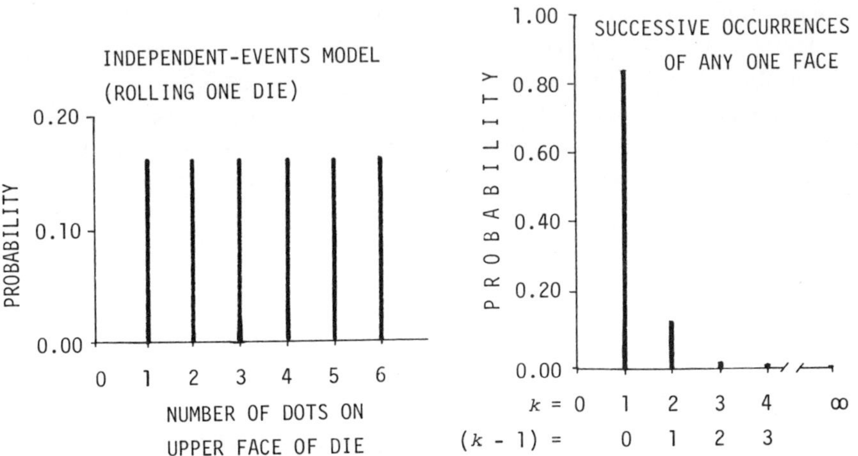

Figure 2. Left, rectangular population density for rolling one six-sided die. Right, geometric population density for runs of 1, 2, 3, . . . successive occurrences of any one face.

bility one-sixth for each face over the range 1 to 6, and zero elsewhere. The histogram displays the total sample space for one die, and it is far from a normal distribution.

Another question of interest in independent-events models is the probability distribution of the same event occurring in successive trials, such as a run of 4s in rolling a die. We ask, "What is the probability that a given side, say 4, will come face up exactly k times in succession, where $k = 1, 2, 3, \ldots?$" We can consider this in terms of a two-state system, in which a 4 face up (state A) prevails until a non-4 face (state B) turns up. The appropriate population density here is the geometric distribution $(1 - p)\,(p)^{k-1}$, where $p = 1/6$, the probability that a 4 occurs on a given toss, and $(1 - p) = 5/6$, the probability that a non-4 occurs on any given trial. The geometric distribution also applies to any single state in a Markov chain, as covered in Fenner (1969, p. WCK–B–13).

For the die example we insert the above values in the density to obtain the probability that state A prevails for exactly k successive rolls, $P\{A = k\} = (5/6)$ $(1/6)^{k-1}$. The histogram in Figure 2, right, shows that the probability of a given face occurring only as a single episode ($k = 1$) is $5/6 = 0.833$; for runs of two ($k = 2$) the probability is 0.138; and for runs of three ($k = 3$) it is 0.023. All runs greater than $k = 3$ have combined probability 0.006. An important characteristic of the geometric distribution is that the mode occurs at $k = 1$. Smart (1968) postulated this distribution for the lengths of stream links, but most of the observational data thus far available suggest a dearth of very short links, just as most observational data on the thickness of rock units in stratigraphic sections suggest a dearth of very thin rock units.

Lengths of stream segments (links) have been considered to be lognormally distributed (Strahler, 1954), though more recently (Smart, 1968; Shreve, 1969; James and Krumbein, 1969) other distributions, based on independent-events models, have been invoked. Similarly, rock-unit thickness is generally taken to be lognormal. These two situations suggest a series of alternatives that deserve detailed study. One is that the generally accepted distributions are correct, in which case some of the stochastic process models may be inappropriate. A second possibility is that the stochastic generating process may be valid, but that subsequent adjustments have "absorbed" the very small individuals by coalescing adjacent nearby tributaries in stream networks and by postdepositional changes in rock units. A third suggestion, which I tend to favor, is that the operational definitions be critically reviewed, in part to see whether the numbers arising from the measurements express the exact geological conceptual meaning behind the operational definitions now used.

Operational Definitions

The identification of first-order stream channels on topographic maps especially is subject to considerable operator judgment. Is part of the problem one of omitting channels too small to show up in the contour pattern? In similar manner, measurement of rock-unit thicknesses, whether as individual beds or as lithologic units, is subject to operator variation in that weathering effects may bias the observations. Even in the subsurface, where micrologs are now abundantly available, considerable judgment is required in deciding exactly where lithologic changes occur.

The question of the geological meaningfulness (and hence usefulness) of numbers generated by particular operational definitions is well illustrated by the concept of shape or form of geological objects. Inasmuch as neither particle shape nor drainage basin shape can be as simply expressed as a set of conic sections, for example, considerable geological judgment is required in developing operational definitions for shape. The examples in Figure 1 include only two of half a dozen basin shape measures, and neither one in fact may be definitive. Any definition is adequate for descriptive purposes, but if basin shape is to be related to processes of basin development, then the exact physical and geological meaning of the numbers becomes highly important, as stressed in connection with Figure 1.

Study of the spatial arrangement of tributaries entering main stream channels (James and Krumbein, 1969) has shown that field conditions in the immediate vicinity of closely spaced tributaries from the same or opposite sides need critical examination; it is likely that the new paths for geomorphic research opened originally by Shreve will place renewed emphasis on field observations, with a much more critical scrutiny of what is to be observed and how it is to be measured.

A second broad area in which operational definitions can advantageously be reviewed is stratigraphic measurement in the field and subsurface records. The literature abounds in measured sections, but many of these are generalized and are of only indirect use in structuring data for Markov models. Subsurface data, especially cores and micrologs, are superior to outcrop data for structuring vertical sequences, but even here operational definitions for picking lithologic changes and measuring individual bed thicknesses need sharpening. Use of digitized logs raises questions of the relative merits of uniformly spaced discrete "probings" of thickness and composition as against continuous measurement. Lateral variations in strata, also important in Markov analysis, are more effectively seen over short distances in the field than in the subsurface, but operational definitions for detecting and measuring cross-cutting relations between time-lines and rock-lines within a limited lateral distance remain to be developed.

CONCLUSIONS

Although only two aspects of quantification in geology—the sharpening of operational definitions and the study of the frequency distributions of the resulting numbers—are emphasized here, other conventional aspects of quantification cannot be neglected. These include the objectivity of the measuring process and the accuracy and precision of the numbers. The burden of evaluating these lies on the developer of the operational definitions. Users of any given operational definition face the problem of deciding whether that definition most nearly expresses the conceptual picture of the quantified attribute in terms of the specific purposes of a particular study. This problem is much less acute in using exemplar methods for measuring lengths, angles, and the like, than it is in measuring more subtle qualities such as shape.

REFERENCES CITED

Anderson, T. W., and Goodman, L. A., 1957, Statistical inference about Markov chains: Annals Math. Statistics, v. 28, p. 89–110.

Fenner, P., ed., 1969, Models of geologic processes: AGI/CEGS short course, Philadelphia, November 1969. Available through Am. Geol. Inst., Washington, D. C.

Griffiths, J. C., 1967, Scientific method in analysis of sediments: New York, McGraw-Hill Book Co.

James, W. R., and Krumbein, W. C., 1969, Frequency distributions of stream link lengths: Jour. Geology, v. 77, p. 544–565.

Krumbein, W. C., 1967, FORTRAN IV computer programs for Markov chain experiments in geology: Kansas Geol. Survey Computer Contr. 13, 38 p.

Krumbein, W. C., and Dacey, M. F., 1969, Markov chains and embedded Markov chains in geology: Math. Geology, v. 1, p. 79–96.

Krumbein, W. C., and Graybill, F. A., 1965, An introduction to statistical models in geology: New York, McGraw-Hill Book Co., 475 p.

Potter, P. E., and Blakely, R. F., 1968, Random processes and lithologic transitions: Jour. Geology, v. 76, p. 154–170.

Shreve, R. L., 1966, Statistical law of stream numbers: Jour. Geology, v. 74, p. 17–37.

—— 1967, Infinite topologically random channel networks: Jour. Geology, v. 75, p. 178–186.

—— 1969, Stream lengths and basin areas in topologically random channel networks: Jour. Geology, v. 77, p. 397–414.

Smart, J. S., 1968, Statistical properties of stream lengths: Water Resources Research, v. 4, p. 1001–1014.

—— 1969, Topological properties of channel networks: Geol. Soc. America Bull., v. 80, p. 1757–1774.

Strahler, A. N., 1954, Statistical analysis in geomorphic research: Jour. Geology, v. 62, p. 1–25.

Watson, R. A., 1969, Explanation and prediction in geology: Jour. Geology, v. 77, p. 488–494.

ADDITIONAL BIBLIOGRAPHY

The references cited have not been updated, but three papers that bear on or expand my continuing work in this area are:

Dacey, M. F., and Krumbein, W. C., 1970, Markovian models in stratigraphic analysis: Math. Geology, v. 2, p. 175–191.

—— 1971, Comments on spatial randomness in dendritic stream channel networks, U. S. Naval Research, Geography Branch, Tech. Rept. no. 17, Contract N00014-67-A0356-0018 (formerly Nonr-1228(36); NR389-150), 18 p.

Krumbein, W. C., 1970, Geological models in transition, *in* Merriam, D. F., ed., Geostatistics: New York, Plenum Press, p. 143–161.

MANUSCRIPT MODIFIED FROM TALK GIVEN AT QUANTITATIVE GEOLOGY SYMPOSIUM AT THE GEOLOGICAL SOCIETY OF AMERICA ANNUAL MEETING IN ATLANTIC CITY, NOVEMBER 10, 1969

Estimating Lithologic Components in Stratigraphic Sequences under Uncertainty

Richard B. McCammon
Department of Geological Sciences
University of Illinois at Chicago Circle
Chicago, Illinois 60680

ABSTRACT

Quantitative estimates of lithologic components in stratigraphic sequences are possible through simultaneous evaluation of acoustic, density, and neutron logs. The interpretative methods developed to determine more accurately the porosity in rocks can be used to determine lithology. In estimating the components for a complex lithology, it is necessary to adopt a probabilistic approach when the number of components exceeds the number of response equations. The maximum-entropy estimate is proposed as the least-biased solution for estimating lithologic components under uncertainty. The estimate is consistent with the information available.

As an example, bed-thickness estimates of idealized cyclical end members are obtained for a sedimentary sequence of Pennsylvanian strata.

INTRODUCTION

For decades geologists have used well logs in subsurface investigations. The information obtained from well logs has provided pertinent data on the lithologic character of subsurface formations, their thickness, and areal distribution. This has resulted in the detailed mapping of the subsurface, providing geologists with a better understanding of recorded geologic history.

A given well log is designed to measure a physical characteristic of the subsur-

face formation being investigated. It records the combined response of formation elements that possess in varying degree the physical characteristic being
measured. A notable advance in recent years has been the development of interpretative logs for the exact determination of the lithic character of subsurface
formations (Savre, 1963; Burke and others, 1967; Dawson-Grove and Palmer
1968; Dorin and Chase, 1968; Harris and McCammon, 1969). Originally these
methods were developed to provide a more accurate estimate of porosity in
rocks and to determine the nature of the contained pore fluids. More recently,
these same methods have been found useful for evaluating mineral quality in
nonmetallic deposits and for estimating properties of rock strength (Tixier and
Alger, 1968; Bond and others, 1969; Jenkins, 1969). The importance of these
developments to stratigraphy is that it is now possible to derive detailed lithologic descriptions of stratigraphic sequences from well logs. In the past such data
could be obtained only by examining drill cores or sample cuttings. Enough
cores are rarely available and sample cuttings recovered during drilling are often
contaminated due to mud-recirculation problems and by hole cavings; however,
well logs, because they are unaffected by these factors, provide a reliable source
of data. Further, a well-log signal penetrates a rock volume greater than the volume represented by the size of the borehole (Lynch, 1962, p. 367) making the
data recovered from well logs a more representative sample of the formation.
Most important, well logs provide quantitative lithologic data that can be
used for improved stratigraphic interpretations. These data can be subjected to
the full spectrum of analytical procedures (Harbaugh and Merriam, 1968). The
use of quantitative variables allows a more formal testing of proposed geological
hypotheses.

However, use of well logs in interpreting lithology has its difficulties. The
specific lithologic character of a sedimentary rock is invariably more complex,
and the changes in lithology with depth more subtle, than well logs now generally can resolve. A rule of thumb for quantitative log evaluation is that the number of logs required to determine the lithic character of a formation is equal to
the number of differentiable lithologic components. This rule places a severe re-
striction on the degree of complexity one can assume in regard to the lithologic
character of a subsurface formation. When, for instance, three logs are available,
we are limited to recognizing three rock types. For alternating strata of different
composition or for a stratum consisting of a complex lithology, quantitative interpretation is difficult. In these cases, we could assume a simple lithology.
However, this quite often leads to meaningless or even spurious interpretations.

Instead, it is possible to accept the complex lithologic nature of layered rocks
and to recognize the uncertainty involved in the situation when the amount of
information available is insufficient to make an unqualified quantitative estimate. The strategy proposed here is to choose a least-biased solution consistent
with the information on hand. To do this, we use the entropy concept as derived from information theory (Shannon, 1948). We consider first how this concept is applied to the determination of porosity for a complex lithology and then
proceed to a stratigraphic example.

POROSITY DETERMINATION

Porosity is the percentage of void space in rocks. In subsurface formations, the void space is invariably fluid-filled and provides a sharp contrast in physical character to the surrounding solid matrix, which has generally a mineral composition. Porosity in the subsurface based on well logs is determined by using one or more of the following three types of logs, each of which measures a different rock property but all of which are strongly influenced by the relative volume of pore fluid contained in the rock.

1. Acoustic log—a recording of interval transit time (Δt) representing the time in microseconds for an acoustic compressional wave to travel one vertical foot of the formation. For many rock types, it has been found experimentally (Wyllie and others, 1956, 1958) that the observed transit time (Δt_o) is related to porosity (ϕ) by the equation

$$\Delta t_o = \Delta t_m (1-\phi) + \Delta t_f \phi \tag{1}$$

where Δt_m represents the interval transit time of the rock matrix and Δt_f represents the interval transit time of the contained pore fluids. The linear relation allows the porosity to be determined if the rock type is known. The fluid velocities are usually given. The matrix characteristics of sedimentary rock types are known from laboratory experiments and field studies.

2. Density log—a recording of the bulk density (ρ_b) in gm/cc of the formation. The bulk density is related to porosity by the equation

$$\rho_b = \rho_m (1-\phi) + \rho_f \phi \tag{2}$$

where ρ_m represents the density of the rock matrix and ρ_f represents the density of the fluid. Again, if the rock type is known, the porosity can be calculated from the observed bulk density. The fluid densities are usually given. The grain density of a mineral is readily determined.

3. Neutron log—a recording of the hydrogen concentration of the formation, thus highly sensitive to water. Because water in addition to filling the pore volume of a rock is also bound to the crystal structure of certain minerals, such as gypsum and various clay minerals, neutron logs are calibrated with respect to a particularly chosen rock type (limestone, sandstone, dolomite) and the neutron response (η) is measured in porosity units of the particular rock type selected. To a first approximation, the observed neutron response (η_o) is related to porosity by the equation

$$\eta_o = \eta_m (1-\phi) + \eta_f \phi \tag{3}$$

where η_m represents the neutron response for 100 percent of the reference rock type and η_f represents the neutron response of the fluid. Neutron logs are generally calibrated so that $\eta_f(100\%) = 1$.

The three logs have this in common: if the rock type is known, the porosity is readily calculable. Further, the response for each log is linear with respect to porosity. Difficulty arises when the lithology has to be determined and is considered to be complex. Clearly, the above equations taken separately are inadequate.

It was recognized several years ago (Savre, 1963) that it was possible to obtain a simultaneous solution in the case of a complex lithology by increasing the number of lithologic components while retaining the linear relations expressed in the equations. To see how this was accomplished, we first simplify our notation. We define x_j as the volume fraction of the jth component for, let us say, a total of n components we wish to consider. We include porosity (ϕ) as one of the volume fractions. Since we have accounted for the total volume, the components must sum to unity, that is,

$$\sum_{j=1}^{n} x_j = 1 \quad . \tag{4}$$

For each component of each log response, we define the response coefficient, a_{ij}, where a_{ij} refers to the response of the jth component for the ith log where the ith log refers to either the acoustic, density, or neutron log. If we designate responses of the three logs y_1, y_2, and y_3, we can write

$$y_i = \sum_{j=1}^{n} a_{ij} x_j \qquad\qquad i = 1, 3 \tag{5}$$

as a system of linear equations that express the observed log response in terms of the volume fractions of the n components we have chosen. The coefficients, a_{ij}, are assumed to be known. We can add equation (4) by specifying $y_4 = 1$ and $a_{ij} = 1$ for all j. Thus, if we have log response data for $(m-1)$ logs, we can express this in the form of m simultaneous linear equations. In order to obtain an exact solution for the given set of simultaneous linear equations and assuming that the log responses are linearly independent, it is well known that m must be equal to n. Thus, if we have data from all three logs, we can specify up to four components. Allowing one component to represent the porosity, we can consider three rock types to be present in varying amounts. Solving the equations, we can determine the exact proportion of each rock type.

If we represent the set of simultaneous equations in matrix form

$$y = \mathbf{A}x \tag{6}$$

where y and x and $m \times 1$ are column vectors and $\mathbf{A}$ is a $m \times n$ matrix, we can write as the solution, when $m = n$,

$$x = \mathbf{A}^{-1} y \tag{7}$$

where $\mathbf{A}^{-1}$ is the inverse of $\mathbf{A}$.

The advantage to be gained by using more than one log to determine porosity is obvious. More components can be considered, allowing more complex lithologies to be interpreted.

Using all three porosity logs, we can treat up to four components (including porosity). This gives much greater flexibility and more versatility in handling the problem of determining porosity in a complex lithology. As it has turned out, however, even greater flexibility is demanded (Burke and others, 1967). The rapid lithologic changes encountered in stratigraphic sequences, the heterogeneous mineral composition found within a single sedimentary rock layer, and the complications due to the complex pore geometries that affect the log response create a situation where it is necessary to consider more components than allowed by the number of logs available. It is true that porosity estimates are not as strongly affected by assuming fewer than what are considered to be the number of components present. However, the limit on the number of components to be considered has marked disadvantages in the interpretation of the lithologic character of formations for stratigraphic purposes.

One solution is to create additional logs, thereby increasing the number of components. However, if such logs are to be useful, they must record a formation characteristic that can be related in a quantitative way to rock type and fluid content. Under certain conditions, logs other than those mentioned meet these requirements. However, the quantitative interpretations of lithology/porosity relations based on these logs is problematical at present so we must seek other solutions. An alternate approach is to recognize the uncertainty inherent in estimating the composition of components in a complex lithology and to obtain a probabilistic solution that has certain desirable statistical properties. Our choice is to select the solution that is least biased but which is consistent with all available information.

PROBABILISTIC APPROACH

Returning to the matrix equation in (6), for $n>m$, there are, in general, an infinite number of solutions. Consider an example involving a three-component system for a single log response, for instance, the value of bulk density recorded on a density log. We assume the system to consist of limestone, dolomite, and fluid-filled porosity. For a given value of bulk density, for example, $\rho_b = 2.67$, we can write

$$2.67 = 2.72 \ x_1 \ + \ 2.86 \ x_2 \ + \ 1.0 \ x_3 \qquad (8)$$

$$1 = \quad x_1 \ + \quad x_2 \ + \quad x_3 \qquad (9)$$

where

$$x_1 = \text{volume fraction of limestone}$$
$$x_2 = \text{volume fraction of dolomite}$$
$$x_3 = \text{fluid-filled porosity} \ .$$

The coefficients in the density equation represent the densities of the different components. For this example, we can display all possible solutions using triangular coordinates in the triangular diagram shown in Figure 1. The part of the line drawn that lies within the triangle contains all possible solutions to the two equations given above and satisfies the added condition that

$$x_j \geqslant 0 \quad \text{for all } j. \tag{10}$$

This insures that solutions containing negative volume fractions will be excluded from consideration. For this case, the possible solutions range from

$$x_1 = .975 \qquad x_2 = 0. \qquad x_3 = .025$$

to

$$x_1 = 0. \qquad x_2 = .90 \qquad x_3 = .10 \ .$$

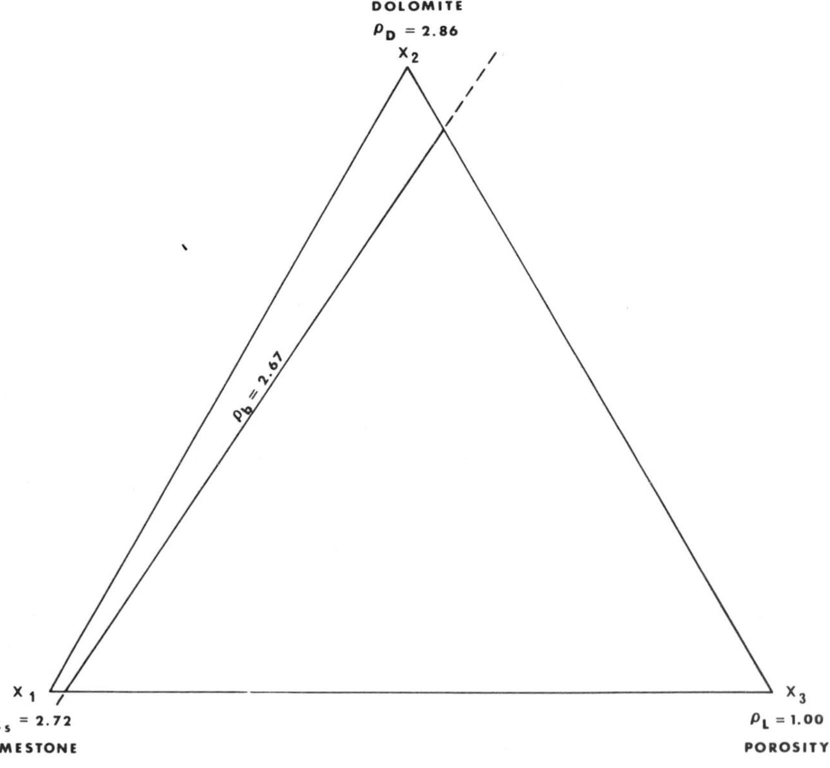

Figure 1. Triangular diagram depicting a three-component system made up of limestone, x_1, $\rho_{L_s} = 2.72$; dolomite, x_2, $\rho_D = 2.86$; water-filled porosity, x_3, $\rho_L = 1.00$. The observed bulk density, ρ_b, equals 2.67 gm/cc.

An infinite number of solutions for components with intermediate mixtures exists. Although the end points provide us with the limits of our uncertainty, we need a criterion for choosing one solution that is best in some sense. Such a criterion necessarily is probabilistic, so we need to adopt a probabilistic approach.

Consider a physical system H made up of a mixture of n components. We assume the components to be randomly interspersed and without mutual interactions. Suppose we wish to measure some property of the system that is possessed in varying degree by every component. Thus, each property can assume only a finite set of values. The number of such values equals the number of components defined for the system. For a system composed of an unknown mixture of n components, the expected value for some measurable property (h) of the system which can take on the values ($h_1, \ldots, h_n$) for the individual components can be expressed as

$$\tilde{h} = \sum_{j=1}^{n} h_j p_j \tag{11}$$

where $\tilde{h}$ is the expected value and p_j is the probability of occurrence for the jth component. For a physical system such as we are considering, each probability can be equated with the volume fraction of each component in the system. The estimation of component fractions is equivalent to the estimate of probabilities.

Our knowledge of the system is based on its measured properties. From these measured properties, we wish to estimate the proportions contributed by the components in the system. As the expression of our knowledge about the system, we use the entropy measure defined from information theory (Shannon, 1948). The entropy of a system with n components is defined by

$$H = -\sum_{j=1}^{n} p_j \ln p_j \tag{12}$$

where H is the entropy and p_j is the probability that the system is composed solely of the jth component. It is a measure of the uncertainty in determining the composition of a mixture of components with unknown probabilities. If we are forced to estimate these probabilities without sufficient information, it has been shown (Jaynes, 1957) that the least-biased solution, that is, the solution giving no particular preference to any component, is given by the maximum of (12) while agreeing with whatever information is given. This has been called (Jaynes, 1957, p. 621) the maximum-entropy estimate; thus, if we have no reason to think otherwise, the probabilities are distributed as evenly as possible. For instance, if we have no information at all, the maximum of (12) occurs for

$$p_j = \frac{1}{n} \qquad \text{for all } j.$$

All the probabilities are equal.

                    RICHARD B. MC CAMMON

Returning to our earlier example, the information we had about the system was the value for bulk density. In this instance, the least-biased solution is obtained when equation (12) is a maximum subject to equations (8) and (9) and the inequalities expressed in (10) have been satisfied. This is shown graphically in Figure 2 where the decile contours for the relative entropy function defined by

$$H = - \frac{100}{\ln(n)} \sum_{j=1}^{n} x_j \ln x_j \qquad (13)$$

are superposed on the triangular diagram. The maximum is found to be

$$x_1 = .43 \qquad x_2 = .50 \qquad x_3 = .07 \quad .$$

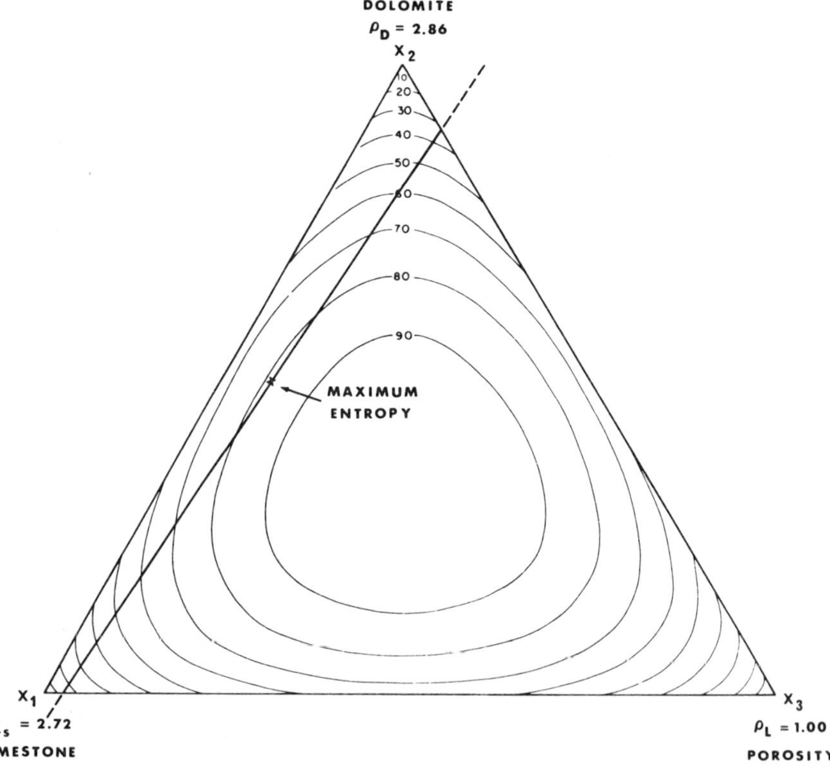

Figure 2. The same as Figure 1 but with the decile contours of the relative entropy function given by

$$H = - \frac{100}{\ln n} \sum_{i=1}^{n} x_i \ln x_i$$

where for this case, $n = 3$, superposed on the triangular diagram. The maximum-entropy solution along the line drawn is

$$x_1 = .43 \qquad x_2 = .50 \qquad x_3 = .07.$$

This represents the least-biased estimate of the proportions of the given components for a mixture having a bulk density of 2.67 gm/cc.

For the case of n components with m equations where n is greater than m, the least-biased solution for estimating the proportions of components in a mixture is given by the maximum for the expression in (12) subject to satisfying the m given simultaneous equations and the nonnegativity constraint. The method of finding the maximum-entropy estimate is described in the Appendix.

The number of components that can be considered now becomes unlimited. We need only take care in our choice of components. In selecting the components to be considered for a complex lithology, we choose those that have a reasonable probability of occurring in the stratigraphic interval in which we are interested.

STRATIGRAPHIC EXAMPLE

Now consider a stratigraphic example involving a complex lithology. Our objective is to reconstruct a lithologic log from the three types of logs discussed previously. For many stratigraphic problems, similar well-log data are available, and we suggest that such logs can be used to provide a basis for establishing lithologic control. As we will see, they can also be used to determine bed thickness.

The example is based on well-log data taken from a published study of a cyclical sedimentary sequence of Pennsylvanian strata in the Illinois Basin (Bond and others, 1969; Jenkins, 1969). These strata have been studied in great detail for several decades. The lithic character, the thickness, and the areal extent of correlative rock stratigraphic units have all been documented in considerable detail (Kosanke and others, 1960). In recent years studies for selected stratigraphic subintervals have resulted in detailed paleoenvironmental mapping of Pennsylvanian cycles (Wanless and others, 1963, 1969; Wanless, 1964). Extensive geologic data collected over several decades have allowed close stratigraphic control.

Outcrop data and examination of thousands of records and samples of subsurface drilling formed the basis of these studies. To improve yet further and eventually to make more quantitative the interpretations based on these data, detailed lithologic interpretations derived from well logs in the manner outlined for porosity determination can be of future value.

Figure 3 is taken from Bond and others (1969) and Jenkins (1969). It represents part of a logged interval for middle Pennsylvanian strata. The strata represented form part of a repetitive sequence consisting of units typical of many cyclothems. What is most often desired in the study of cyclothems is to identify the rock units that make up the ideal cyclothem (Kosanke and others, 1960, Fig. 2) and to determine the thickness and areal extent of each rock unit. For this example, we are interested in identifying the units present and in determining the thickness of each unit based on the average log characteristics. The selected interval extends from the lower boundary of the bottom coal to the contact with the upper coal. This comprises one cycle of the sequence. The average values over the interval for the three logs, that is, the acoustic, density, and neutron logs, were calculated as 114.7 μsec/ft, 2.13 gm/cc, and 0.48 limestone por-

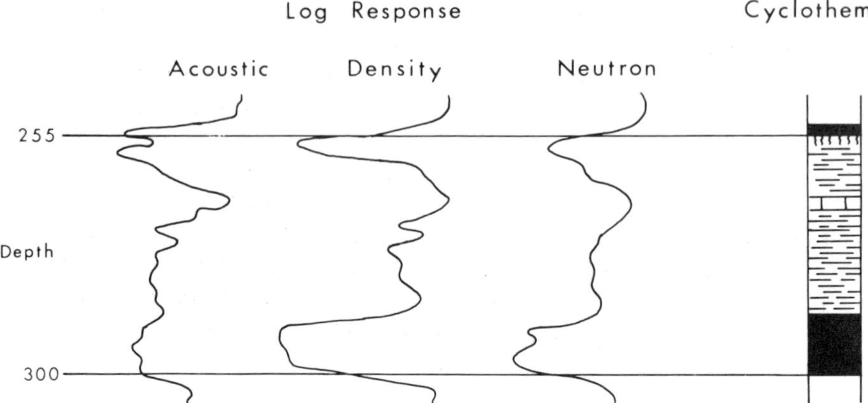

Figure 3. Part of the stratigraphic interval of middle Pennsylvanian strata taken from Bond and others (1969, Fig. 1). The response of the acoustic, density, and neutron logs are shown opposite the graphic log obtained from core descriptions.

osity units. Six rock types were selected to represent the lithic character of the interval. The response coefficients for these components are given in Table 1. Thus, we have four equations with six unknowns. We wish to obtain a solution consistent with this information. The preferred solution as we have shown is the maximum-entropy estimate and it is given in Table 2. The maximum-entropy estimate of bed thicknesses can be compared with the thicknesses determined by inspection from the recovered cores. These thicknesses are given in the right-hand column of Table 2. The estimates based on maximum entropy are roughly comparable to the core determined thicknesses. Differences do exist, however. Some of the rock types chosen are in fact mixtures of purer rock types with re-

**Table 1. Rock-Unit Logging Parameters for
Stratigraphic Interval Shown in Figure 3**

	Log type		
Rock unit	Acoustic Δt μ sec/ft	Density ρ_b gm/cc	Neutron η 1s porosity units
limestone	50	2.70	0.00
sandstone	55	2.50	0.10
siltstone-shale	90	2.45	0.20
black shale	110	2.20	0.30
gray shale	120	2.35	0.40
coal	130	1.40	1.00

gard to texture and composition, so it is unlikely that the visual estimates of unit thickness should correspond to calculated thickness. For instance, the core description included some sandy and silty black shales which were calculated partly as siltstone-shale and sandstone based on the maximum entropy solution. A second difference is that the impurities in the coal as seen from the logs had the effect of reducing the estimated coal thickness while increasing the thickness of the impurity component. This suggests that the maximum-entropy solution provides estimates of the net thickness of the lithologic components that compose the sequence. Unquestionably more work needs to be done in determining the well-log characteristics of specific rock types. However, as this example shows, it is possible to derive more than just gross lithology from well logs.

There is little doubt that in the future quantitative lithologic interpretations from logs will be made on a routine basis. When this happens, a significant step will have been taken in the further quantification of stratigraphy.

ACKNOWLEDGMENTS

The idea that the maximum-entropy concept could be useful for estimating lithologic components under uncertainty was first suggested to me by J. E. Warren of Gulf Research and Development Company, where much of the early work on the problem was carried out. Subsequent development of the method to obtain a more accurate estimate of porosity using well logs was carried out in collaboration with M. H. Harris from the same company. The material presented here represents a further extension of that work.

Table 2. Maximum Entropy-Derived Estimates of Rock Unit Thicknesses for Stratigraphic Interval Shown in Figure 3

	Estimated thickness (ft)	Core measured thickness (ft)
siltstone-shale	4	0
limestone	1	2
black shale	9	15
gray shale	20	16
coal	10	11
underclay	..	<1
sandstone	1	0
Total thickness	45	45

APPENDIX.   OBTAINING THE MAXIMUM-ENTROPY ESTIMATE

Consider that we have a set of m simultaneous equations in n unknowns $(x_1, \ldots, x_n)$ given by

$$y_i = \sum_{j=1}^{n} a_{ij} x_j \qquad i = 1, \ldots, m$$

where the y_i's and the a_{ij}'s are known. We have further that $y_m = a_{mj} = 1$ so that

$$\sum_{j=1}^{n} x_j = 1.$$

For $n > m$, there is, in general, no exact solution for $(x_1, \ldots, x_n)$. Rather, an infinite number of solutions satisfy the above equations. For our purpose, we wish to find the solution for which the entropy expression defined by

$$H = - \sum_{j=1}^{n} x_j \ln x_j$$

is a maximum subject to

$$x_j \geqslant 0$$

for all j. This solution is called the maximum-entropy estimate (Jaynes, 1957).

To find the solution, we introduce Lagrange multipliers $(\lambda_1, \ldots, \lambda_m)$ and define a function

$$F = - \sum_{j=1}^{n} x_j \ln x_j + \sum_{i=1}^{m} \lambda_i \left(\sum_{j=1}^{n} a_{ij} x_j - y_i \right)$$

which we maximize unconditionally. To find the maximum, we set the partial derivatives

$$\frac{\partial F}{\partial x_j} = - \ln x_j - 1 + \sum_{i=1}^{m} \lambda_i a_{ij} \qquad j = 1, \ldots, n$$

equal to zero and solving for x_j, we obtain

$$x_j = e^{\sum_{i=1}^{m} \lambda_i a_{ij} - 1} \qquad j = 1, \ldots, n .$$

Whatever the values we obtain for the λ_i's, we will have that

$$x_j \geqslant 0 \qquad \text{for all } j$$

thus satisfying the nonnegativity constraint.

To solve for the values of λ_i, we substitute the expression for x_j into the original equations to obtain

$$y_i = \sum_{j=1}^{n} a_{ij}\, e^{\sum_{k=1}^{m} \lambda_k a_{kj} - 1} \qquad i=1,\dots,m$$

which gives us m nonlinear equations with respect to the m unknown λ_k's. The values of the λ_k's can be found by iteration using nonlinear methods. A computer program for solving nonlinear equations is available (McCammon, 1969). After finding the values of the λ_k's, these can be substituted into the equations for the x_j's which become the maximum-entropy estimate.

REFERENCES CITED

Bond, L. O., Alger, R. P., and Schmidt, A. W., 1969, Well log applications in coal mining and rock mechanics: AIME preprint 69–F–13, 19 p.

Burke, J. A., Curtis, M. R., and Cox, J. T., 1967, Computer processing of log data improves production in Chaveroo Field: Jour. Petroleum Technology, v. 19, p. 889–895.

Dawson-Grove, G. E., and Palmer, K. R., 1968, A practical approach to analysis of logs by computer: Second Formation Evaluation Symposium, Canadian Well Logging Soc., 12 p.

Dorin, A. H., and Chase, A. E., 1968, A guide to improved formation evaluation in carbonates of northwestern Alberta: Second Formation Evaluation Symposium, Canadian Well Logging Soc., 10 p.

Harbaugh, J. W., and Merriam, D. F., 1968, Computer applications in stratigraphic analysis: New York, John Wiley & Sons, 282 p.

Harris, M. H., and McCammon, R. B., 1969, A computer oriented generalized porosity-lithology interpretation of neutron, density, and sonic logs: AIME preprint SPE 2528, 16 p.

Jaynes, E. T., 1957, Information theory and statistical mechanics: Phys. Rev., v. 106, p. 620–630.

Jenkins, J. C., 1969, Practical applications of well logging to mine design: AIME preprint 69–F–73, 20 p.

Kosanke, R. M., Simon, J. A., Wanless, H. R., and Willman, H. B., 1960, Classification of the Pennsylvanian strata of Illinois: Illinois Geol. Survey Rept. Inv. 214, 84 p.

Lynch, E. J., 1962, Formation evaluation: New York, Harper & Row, 422 p.

INTRODUCTION

Geologic quantification has tended to develop an unfortunate credibility gap or communication barrier between field-oriented geologists and the emerging group of instrument-analysis, computerized, geostatistical scientists. Part of the sometimes near hostility stems from a feeling of obsolescence on the part of scientists trained prior to 1960, before statistical instruction became a common part of geologic training. Many younger geologists, on the other hand, have been so occupied acquiring quantitative skills that they have missed the opportunity to get acquainted with rocks in the field. Some recently trained geoscientists spent less than 60 days doing field work before acquiring the Ph.D. They may fail to appreciate the vast amount of descriptive field work that is still needed to characterize adequately the lithology, age, and structures of the rocks at the earth's surface.

Unfortunately, field workers are plagued by discontinuous exposures and by lack of time to record all observable aspects of the rocks. These characteristics of field geology lead to uncertainties of sampling from a total population, so that the results must be scrutinized statistically to evaluate reliability of field conclusions and to search for geological patterns.

My own research experience has been chiefly interpreting stratigraphic and structural field relations of Appalachian Paleozoic sediments. Because statistical training may lead to a probabilistic view of one's total existence, I want to point out some statistical implications of stratigraphic field work that seldom are recognized by workers studying stratigraphic field relations.

One thing is certain: statistical aspects of rocks in the field will be seen only if the geologist has specific statistical training. Almost without exception, attempts to analyze statistically field observations collected by the nonstatistical scientist reveal that statistical planning before going in the field would have greatly enhanced interpretation of the results. To obtain maximum geological meaning from field work, one must do the field work using a statistical sampling plan, and then analyze and interpret the results.

MAP RESOLUTION

A geological map portrays for a total region the areal distribution of some geological variable, such as rock types recognizable by carefully walking out the formation contacts over an entire area. Formations are seldom completely exposed, and the geologist rarely has time to examine the entire area in great detail. Most geologic maps are actually interpretations of outcrop maps that portray the rocks really seen by the geologist (Kupfer, 1966). Even an outcrop map is somewhat interpretative, because the geologist must assign the rocks of an outcrop to specific formations. Published geologic maps can almost always be improved by more detailed field studies.

Increasingly, geologists make maps based on interpretations between sample-control points. This is true of a map showing paleogeology beneath an unconformity, based on data from outcrop and well-control points. Isopachous maps provide another example. Paleogeology and stratigraphic thickness at least can be visualized in the field at control stations. Modern isopleth maps of trace-element distribution, such as chemical halos around ore bodies, are clearly based on control points for which the mapped data are totally invisible in the field.

Map reliability and resolution vitally concern all geoscientists. It is commonly recognized that improved maps result from more detail in field studies and closer spacing of control points; nevertheless, geologic literature contains little evaluation of this problem. Muehlberger and others (1967, p. 2354–2362) faced the problem as they attempted to develop a basement map for the interior of the United States. Their work illustrated the increased resolution that resulted from better control by comparing the detail discernible from scattered well data with the complexity of similar Precambrian terrain as mapped in outcrop areas of the Llano Uplift and Black Hills.

Dodd and others (1965) and Cain (1966) vividly illustrated the human tendency to overinterpret data. They generated a random distribution of numbers on a map grid and found that geologists tend to interpret geological meaning from truly random patterns. We geologists are often so eager to find order and patterns in nature that we read into our data far more meaning than is justified.

The following example illustrates semiquantitatively the sort of increasing resolution that is possible with improving geologic sample control (Fig. 1). An aeronautical chart (Meyers, 1965) showing the lower Mississippi Delta with a scale of 1:500,000 was used to establish the true distribution of three sedimentary environments (land, sea, and fresh water) in a square 78 miles on a side. Random-number tables were used to select sets of 12, 25, 60, and 300 "sample sites" superimposed on the aeronautical chart. A special set of 169 control points was prepared by laying a 6-mile square grid corresponding to townships over the map, and then using a random-number table to locate one random control site within each township. Various geologists then constructed environment maps from the data at these control points. Increasing resolution with increasing control density is apparent. Many published stratigraphic environmental interpretation maps come from control densities comparable to the maps with 12 or 25 data points.

The 169-point map is particularly interesting, because it represents a control density of one point per township. This is the control density desired in the Pennsylvanian environmental maps developed by Wanless and others (1963, p. 437), maps generally considered among the best regional stratigraphic maps ever constructed. Yet nearly twice that control density is necessary to reveal the true complexity of the Mississippi Delta and its associated barrier islands.

In this example there is no difficulty in resolving geologic time. With the added complication of correctly interpreting temporal relations among the stratigraphic columns, the reliability of most paleoenvironmental maps is clearly much less than generally assumed.

As a class exercise I have asked students to determine random points on a topographic map, mark the elevation of those points, and then have other geology graduate students innocently prepare contour maps from the data. Re-

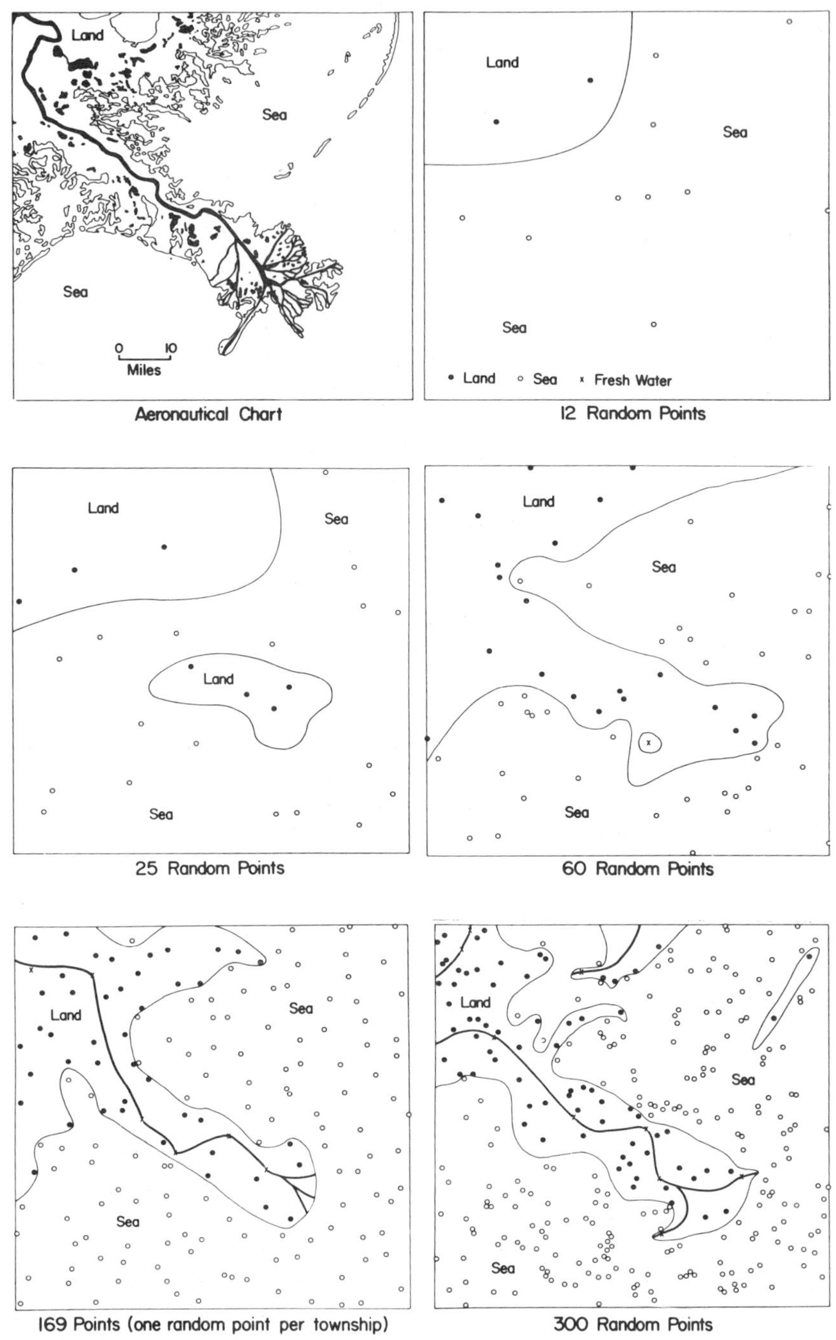

Figure 1. Aeronautical chart of Mississippi Delta is sampled with random points to show improving resolution of three environments as the number of sample points increases.

markable lack of resolution sometimes occurs: the Grand Canyon trends the wrong way or the Allegheny Front is not detected until a large number of points is obtained. The implications for interpreting samples of the invisible data, such as geochemical samples, are impressive.

PRECISION OF STRATIGRAPHIC THICKNESS

For stratigraphers, a basic item of data is the thickness of a particular formation or other stratigraphic unit. Assuming that the formation boundary can be designated without error, the stratigrapher must still face the problem of reproducibility of thickness determinations, a problem about which there is little information available. All measurements, even those made by dropping a tape over a cliff in horizontal beds, have some statistical error in their evaluation. Thickness of dipping strata measured along a sloping traverse oblique to the strike is somewhat less reliable—whether determined by Brunton-compass and tape or by plane table. Surely all stratigraphers agree that interpretations or explanations of thickness differences are meaningless if the difference between two outcrops is less than the error of measurement.

Stratigraphers seldom concern themselves with the precision of their stratigraphic sections, commonly measuring a section only once and recording its thickness as fact rather than as a best estimate. This avoids the problem of reliability but also makes calculation of precision impossible. (How reliable is an isopachous map if the precision of the individual control points is unknown?)

The following paragraphs describe calculations of my own precision in measuring stratigraphic thickness, measurements that are probably typical of those made by stratigraphers.

I made duplicate measurements of the same section, permitting calculation of thickness precision, for eight beds in each of four structural settings (Table 1). Error was calculated in decimal parts of total bed thickness, using the following procedures.

$$X'_1 = \frac{X_1}{\dfrac{X_1 + X_2}{2}} \qquad\qquad X'_2 = \frac{X_2}{\dfrac{X_1 + X_2}{2}}$$

$$D = X'_1 - X'_2$$

For the data from the eight pairs of readings with $0°$ to $30°$ dip amount and $0°$ to $30°$ between traverse and strike, the precision of individual thickness measurements is calculated as follows:

**Table 1. Brunton-Compass and Tape-Thickness Measurements
and Summary of Calculations for Standard Deviation
of Individual Thickness Measurements**

		Data pair number	X_1 (feet)	X_2	X'_1	X'_2	$D = X'_2 - X'_2$	
$0°-30°$ between traverse and strike	$0°-30°$ dip	1	16.7	16.3	1.012	.988	.024	
		2	41.2	41.4	.998	1.002	−.004	
		3	7.0	6.9	1.007	.993	.014	
		4	10.54	9.73	1.040	.960	.080	$s_{X'} = .0348$
		5	20.3	21.0	.983	1.017	−.034	
		6	9.4	9.3	1.005	.995	.010	
		7	43.3	47.5	.954	1.046	−.092	
		8	23.9	24.0	.998	1.002	−.004	
	$30°-90°$ dip	9	19.3	20.1	.980	1.020	−.040	
		10	7.40	7.36	1.003	.997	.006	
		11	14.2	15.9	.944	1.056	−.112	
		12	13.26	12.40	1.034	.966	.068	$s_{X'} = .0675$
		13	24.0	24.4	.992	1.008	−.016	
		14	11.25	10.36	1.041	.959	.082	
		15	8.43	10.29	.901	1.099	−.198	
		16	25.2	28.0	.947	1.053	−.106	
$30°-90°$ between traverse and strike	$0°-30°$ dip	17	3.32	3.98	.910	1.090	−.180	
		18	32.3	30.8	1.024	.976	.048	
		19	24.9	24.5	1.008	.992	.016	
		20	36.3	33.2	1.045	.955	.090	$s_{X'} = .0628$
		21	8.99	9.60	.967	1.033	−.066	
		22	13.5	14.1	.978	1.022	−.044	
		23	15.9	16.8	.972	1.028	−.056	
		24	25.7	24.0	1.034	.966	.068	
	$30°-90°$ dip	25	49.6	49.3	1.003	.997	.006	
		26	28.3	29.8	.974	1.026	−.052	
		27	31.3	33.7	.963	1.037	−.074	
		28	33.7	33.8	.999	1.001	−.002	$s_{X'} = .0465$
		29	10.8	10.7	1.005	.995	.010	
		30	26.1	22.5	1.074	.926	.148	
		31	44.9	44.1	1.009	.991	.018	
		32	49.2	49.5	.997	1.003	−.006	

for all 32 data pairs $s_{X'} = .0532$

$$s_D = \sqrt{\dfrac{\Sigma D^2 - \dfrac{(\Sigma D)^2}{8}}{7}} = \sqrt{\dfrac{.016924 - \dfrac{(-.006)^2}{8}}{7}} = .0492$$

$$s_{X'} = \dfrac{s_D}{\sqrt{2}} = .0348$$

Thus the standard deviation of individual thickness measurements is 3.48 percent.

At the right of Table 1 is a summary of the standard deviations for each set of eight data pairs. The pooled standard deviation of a typical Brunton-compass and tape thickness measurement is 5.32 percent, calculated as follows from all 32 data pairs:

$$s_D = \sqrt{\dfrac{\Sigma D^2 - \dfrac{(\Sigma D)^2}{32}}{31}} = \sqrt{\dfrac{.180844 - \dfrac{(-.398)^2}{32}}{31}} = .0753$$

$$s_{X'} = \dfrac{s_D}{\sqrt{2}} = .0532$$

This provides a measure of the precision of typical thickness determinations. Assuming that measuring errors are normally distributed, 95 percent confidence limits for the precision of a single thickness measurement are $\pm (2.02)\ (.0532)$, that is, about 11 percent.

Preliminary investigations of the precision of thickness measurements of strata by plane-table techniques indicate that the plane-table procedure is not significantly better than the Brunton-compass and tape method. This is apparently because strike and dip determination is the most variable element in both procedures, and it overshadows other sources of error. In certain field situations, plane-table methods are easier to execute than Brunton-compass and tape traverses, but use of the plane-table apparently offers no significant improvement in reliability of results.

Preparation of isopachous maps involves comparing one thickness with thickness values for several surrounding points. Duncan's new multiple range test (Duncan, 1955; Steel and Torrie, 1960, p. 107–109, 442–443) is considered appropriate for determining the least contour difference that can be detected with 95 percent certainty. Assuming that one point is compared with four other points for contouring (comparison of a total of five points), the least significant range that can be detected, according to my calculations as presented above, is 3.12 standard deviations, or 16.6 percent of the total thickness. This implies, of course, that different contour spacings should be used for different parts of the map. Based on my field data, the contour values that show the maximum detail of reliable information (*see* Fig. 2) are as follows: 1, 2, 3, 4, 5, 6, 7, 8, 10, 12, 14, 17, 20, 24, 28, 33, 39, 45, 53, 62, 72, 85, 100, 118, 140, 175, 205, 240, 280, 330, 385, 450, 530, 620, 720, 850, and 1,000 feet. Conventional isopachous maps with uniform isopach intervals are not uniformly reliable over the entire map area. A conventional uniform contour spacing of 10 feet discards much reliable detail in strata 0 to 60 feet thick, yet implies a greater knowledge than is reliable for regions with a stratigraphic thickness of more than 100 feet.

The variable isopach spacing in Figure 2 is used to portray the thickness of the Devonian Onesquethaw Stage in parts of Virginia and West Virginia. The Onesquethaw Stage was defined (Dennison, 1961, p. 10) to include those strata from the top of the Oriskany Sandstone and its correlates to the top of the Tioga Bentonite. The stratigraphy of the Tioga tuffaceous beds is now well enough understood to know that the "middle coarse zone" of tuff within the Tioga of Virginia is the stratum recognized as the Tioga Bentonite in the standard Devonian section of New York, so the position of the "middle coarse zone" of the Tioga is now used to mark the top of the Onesquethaw Stage (Dennison, 1969). The basal contact of the Onesquethaw Stage is very abrupt in the map area and is unconformable over much of the region. Therefore the interval to be isopached is clearly defined, and the principal difficulty is precision of stratigraphic thickness measurements. The map in Figure 2 shows the maximum amount of reliable information. It also represents a substantial revision, based on new thickness data, of an earlier isopachous map of the Onesquethaw Stage (Dennison, 1961, p. 12).

This variable isopach interval procedure has several advantages:

1. It portrays the maximum amount of reliable geologic information.
2. The map has a uniform reliability throughout its extent.
3. The level of reliability is known.

It also has disadvantages:

1. Greater care must be used to interpolate the position of isopachs between control points.
2. Rate of thickness change is not readily apparent. With uniform isopach interval, a change in map spacing of isopachs indicates a change of gradient. This would not be true for the proposed variable contour procedure. Uniform rate of thinning is represented by varying isopach spacing, and a rare appearance of uniform map spacing of isopach indicates a specific and peculiar change in rate of thinning.

In my opinion, the advantages justify using the variable isopach spacing procedure, although it means modifying several customary habits of geological map interpretation.

In the northwest part of Figure 2, thickness values are based on well control, and true thickness is known somewhat more precisely than for the Brunton-compass and tape-thickness data (true thickness from well cuttings is known there with less than 10 feet of error, in my judgment). It may be appropriate to deviate from the contour spacing suggested for outcrop thickness precision, using a second method of fixed isopleth spacing for subsurface data. Interpreta-

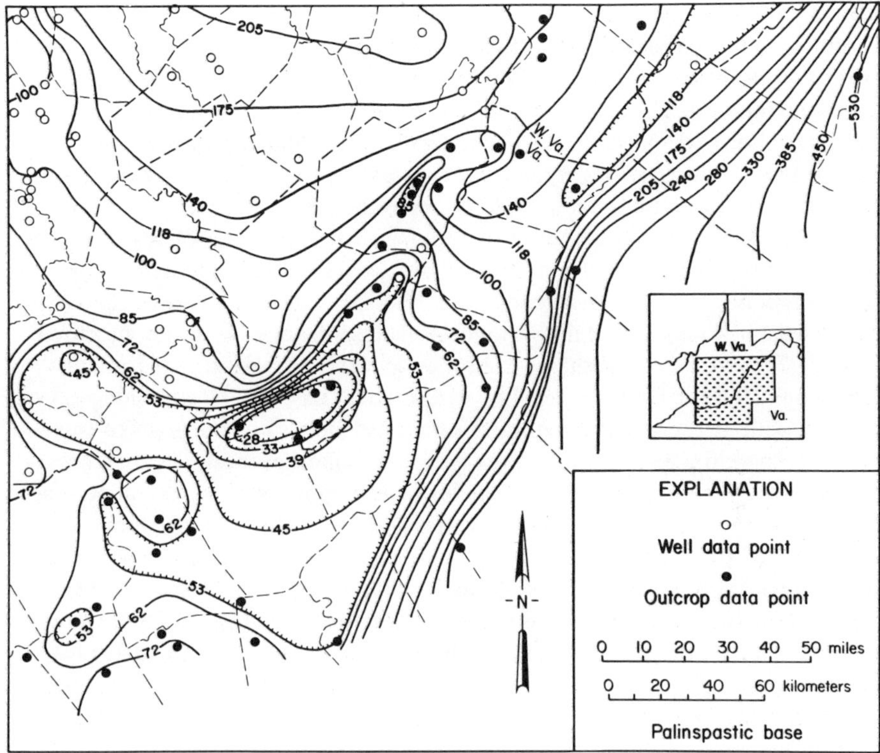

Figure 2. Isopachous map (in feet) of Devonian Onesquethaw Stage for part of Virginia and West Virginia. Note variable isopach interval used to portray maximum amount of uniformly reliable thickness information.

tion of such a map with two isopach procedures may be confusing, however. Consequently, the variable contour interval for outcrop data is used in the subsurface control part of the map for consistency throughout the map area, even though not all the usable subsurface information is portrayed. By analogy, the total chain is as strong as its weakest link.

STATISTICAL FEEDBACK IN FIELD WORK

A working knowledge of statistical techniques is desirable before field sampling begins, but experience gained in the field can also suggest improvements and modifications of the sampling design. For maximum extraction of statistical meaning, there should be feedback between the statistical and field phases of the work.

For several years I have been studying the Devonian delta stratigraphy along the Allegheny Front from Maryland to Virginia. Only three measured sections had been described previously in that general area (summarized in plates C and D, Woodward, 1943). From these sections (Fig. 3) no regional trends in sedimentation traits along the outcrop belt could be discerned in the Upper

Devonian Chemung Formation. At the outset of my work the most urgent need was simply a detailed description of the strata.

Field work finally resulted in ten measured sections of the Chemung Formation in the outcrop belt. This presented an opportunity to study statistical trends in sedimentation traits. As one example of a sedimentary parameter that can be analyzed statistically, let us consider grain-size trends in sandstone. For each thin-section sample we could measure the maximum diameter of the ten largest grains and obtain a mean size value for these grains. Random-number tables could be used to select a stratigraphic position for a sample within the measured section, and for each point the closest sandstone could be sampled. Thus it would be easy to obtain six random samples from each of ten measured sections. This would establish control for an analysis of variance experiment with a completely random design (Fig. 4). It would have a total of 59 degrees of freedom, with 9 degrees of freedom for sections and 50 error degrees of freedom. This would permit an objective test of the null hypothesis that in all measured sections the sandstones have a uniform mean grain size for maximum diameter of the ten largest grains. Further, the sections' degrees of freedom could be partitioned for a regression test of grain-size change with distance along the outcrop belt. This plan permits systematic study of grain sizes with the Chemung Formation. (Decrease in numbers of sandstone beds toward the southwest leads me to speculate that the grain-size mean of sandstones also decreases toward the southwest.)

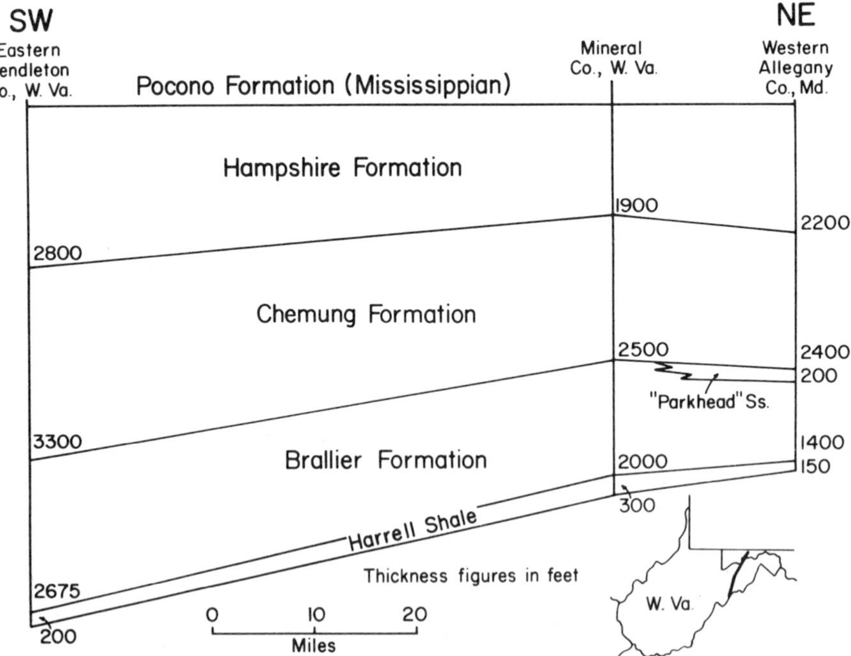

Figure 3. Upper Devonian stratigraphy along part of Allegheny Front as understood before 1960 (after Woodward, 1943).

At this point very detailed field work indicated the desirability of stratigraphic refinement and also the possibility of a more sophisticated statistical analysis. The Chemung Formation of earlier workers is readily divisible into two mappable units (Dennison, 1963) that in Figure 5 are labeled Formations I and II. Further, Formation II can be divided into five members. Formal revision of this stratigraphic nomenclature was recently published (Dennison, 1970), along with the stratigraphic cross section of Figure 6. Grain size of Formation I can be sampled best with a completely random design, using five samples from each of nine stratigraphic sections, as shown in Figure 5. Formation II can be sampled by a factorial design (Cochran and Cox, 1957, p. 148–182; Steel and Torrie, 1960, p. 194–231). The factorial design permits testing for grain-size differences among stratigraphic sections and among members, plus testing for interaction between sections and members.

In this factorial design it would be possible to partition degrees of freedom for sections to test by regression for a grain-size trend with distance along the line of sections. It is obvious in the field that sandstone members B and D are markedly coarser on the average than the mixed siltstone, sandstone, and shale which comprise members A, C, and E, hence obvious choices for partitioning the 4 degrees of freedom for members would be to make these comparisons: B and D versus A, C, and E; B versus D; A versus C and E; and C versus E. Stratigraphic and sedimentologic analysis suggests that member A is chiefly below wave-

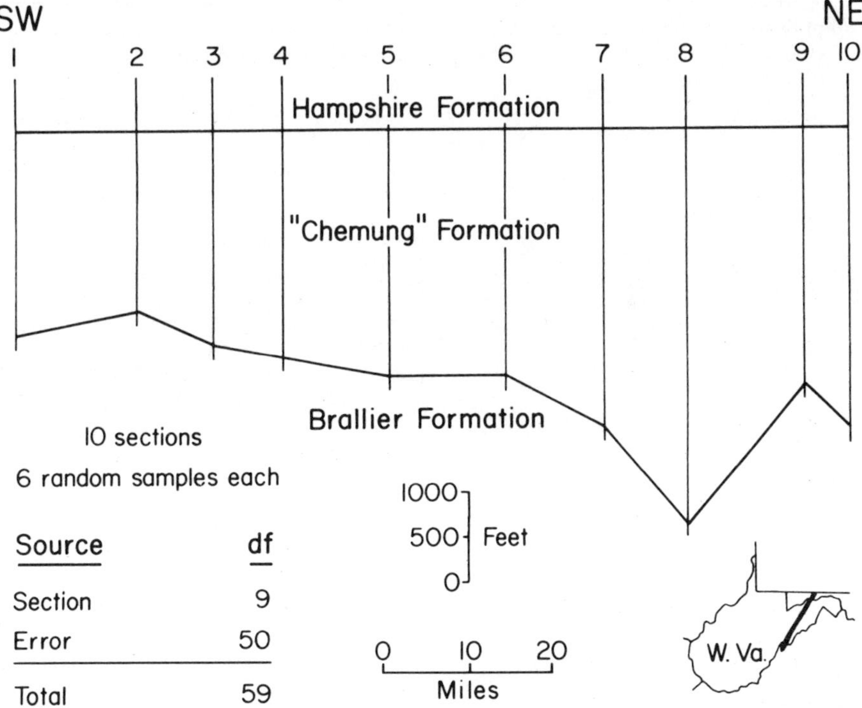

Figure 4. Completely random design to study grain-size variation in Chemung Formation along Allegheny Front.

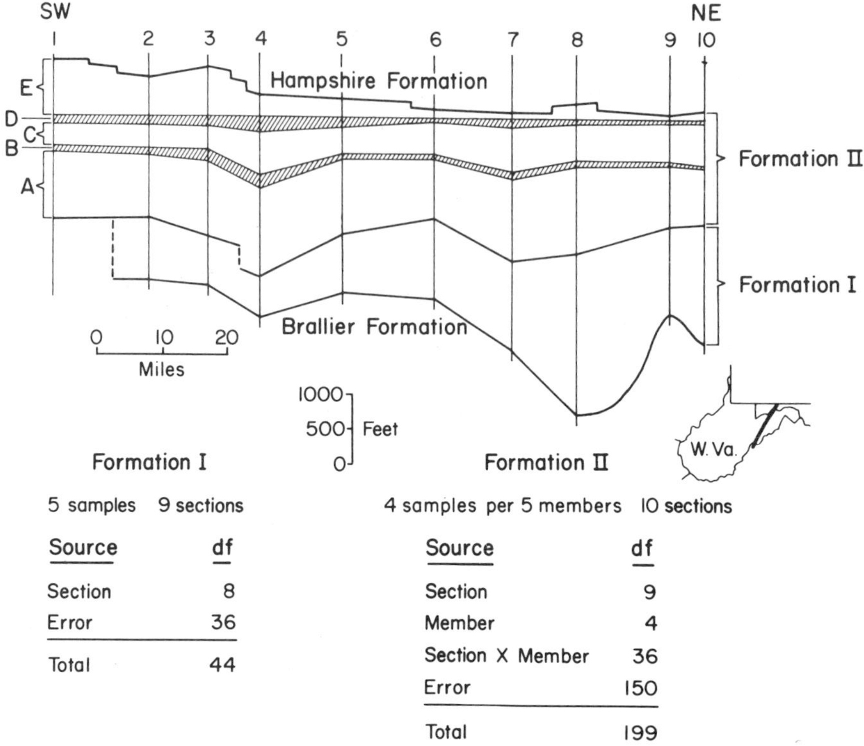

Source	df
Section	8
Error	36
Total	44

Source	df
Section	9
Member	4
Section X Member	36
Error	150
Total	199

Figure 5. Final experimental design to study grain-size variation in two new formations within the old Chemung Formation mapping unit along Allegheny Front. Formation I is analyzed with a completely random design. Formation II is sampled using a factorial design.

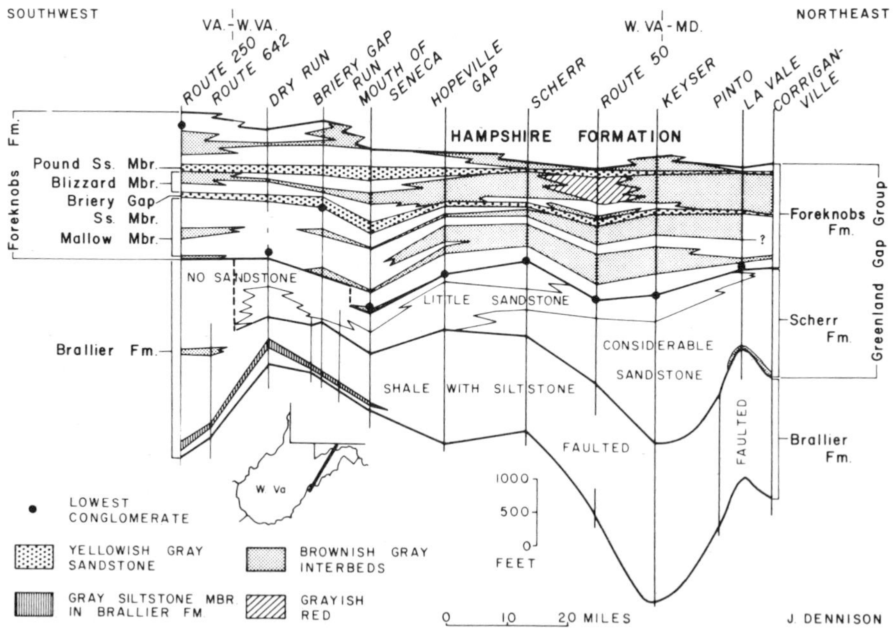

Figure 6. Current stratigraphic nomenclature and facies interpretation of Upper Devonian marine strata along Allegheny Front (from Dennison, 1970).

base, while members C and E are nearshore and shallow, brackish sediments. Members B and D are winnowed marine sediments, probably barrier or possibly delta front sands.

A factorial analysis of stratigraphic variables was described by Krumbein (1955), who cited an example of map grid factors, rather than factors in the vertical plane of a stratigraphic cross section.

The final statistical design of Figure 5 could not have been anticipated before detailed field studies suggested the presence of formations and members within what was originally called the Chemung Formation. Strict adherence to a sampling pattern developed before the phase of detailed field studies would not have permitted answering so many and such rigorous questions.

The above example vividly illustrates the general conclusion that geologists should always remain alert for statistical feedback during field work.

REFERENCES CITED

Cain, J. A., 1966, Further random thoughts on randomly located random numbers: Jour. Geol. Education, v. 14, p. 62–63.

Cochran, W. G., and Cox, G. M., 1957, Experimental designs: New York, John Wiley & Sons, 2d ed., 611 p.

Dennison, J. M., 1961, Stratigraphy of Onesquethaw Stage of Devonian in West Virginia and bordering states: West Virginia Geol. Survey Bull., v. 22, 87 p.

—— 1963, Stratigraphy between Tioga Bentonite and Pocono Formation along Allegheny Front in Maryland and West Virginia: Pennsylvania Topographic and Geol. Survey, General Geol. Rept. G39, p. 213–227.

—— 1969, Tioga Bentonite and other isochronous units in the Devonian Oriskany to Tully interval of the Appalachian Basin: Geol. Soc. America, Abs. with Programs for 1969, Pt. 1 (Northeastern Section), p. 13.

—— 1970, Stratigraphic divisions of Upper Devonian Greenland Gap Group ("Chemung Formation") along Allegheny Front in West Virginia, Maryland, and Highland County, Virginia: Southeastern Geology, v. 12, no. 1, p. 53–82.

Dodd, J. R., Cain, J. A., and Bugh, J. E., 1965, Apparently significant contour patterns demonstrated with random data: Jour. Geol. Education, v. 13, p. 109–112.

Duncan, D. B., 1955, Multiple range and multiple F tests: Biometrics, v. 11, p. 1–42.

Krumbein, W. C., 1955, Statistical analysis of facies maps: Jour. Geology, v. 63, p. 452–470.

Kupfer, D. H., 1966, Accuracy in geologic maps: Geotimes, v. 10, no. 7, p. 11–14.

Meyers, H. B., 1965, Official aeronautical chart of Louisiana: Louisiana Dept. Public Works.

Muehlberger, W. R., Denison, R. E., and Lidiak, E. G., 1967, Basement rocks in continental interior of United States: Am. Assoc. Petroleum Geologists Bull., v. 51, p. 2351–2380.

Steel, R.G.D., and Torrie, J. H., 1960, Principles and procedures of statistics
    with special reference to the biological sciences: New York, McGraw-Hill
    Book Co., 481 p.
Wanless, H. R., Tubb, J. B., Gednetz, D. E., and Weiner, J. L., 1963, Mapping
    sedimentary environments of Pennsylvanian cycles: Geol. Soc. America
    Bull., v. 74, p. 437–486.
Woodward, H. P., 1943, Devonian System of West Virginia: West Virginia Geol.
    Survey Bull., v. 15, 655 p.

MANUSCRIPT MODIFIED FROM TALK GIVEN AT QUANTITATIVE GEOLOGY SYM-
POSIUM AT THE GEOLOGICAL SOCIETY OF AMERICA ANNUAL MEETING IN
ATLANTIC CITY, NOVEMBER 10, 1969

The Geological Society of America, Inc.
Special Paper 146, © 1972

Trend Surface Analysis
and Spatial Correlation

Geoffrey S. Watson
Department of Statistics
Princeton University
Princeton, New Jersey 08540

ABSTRACT

The methods for analyzing data in one dimension (usually time) are highly developed. In several dimensions, however, most applied workers use only one class of methods: that in which the data are assumed to be the sum of a deterministic trend and an uncorrelated random error. The more general model with a spatially correlated error has been neglected in most fields, oceanography being a conspicuous exception. This neglect is due partly to the lack of expositions for the applied workers. However, the manner in which much geological data are now collected does make this application difficult.

INTRODUCTION

Mapping has always been a favorite tool of geologists so that, with the increased accessibility of computers, it is scarcely surprising that geologists should now be very interested in studying spatial distributions mathematically. Their aim is to use some mathematical model to smooth or contour a net of data points, or to interpolate between them, or to calculate some average over the area (Harbaugh and Merriam, 1968; Merriam and Cocke, 1967).

There are several possible methods to achieve this aim, each with its advocates and detractors. In the United States the most popular methods are based on the assumption that

$$value\ at\ data\ point\ =\ value\ of\ deterministic\ function + random\ error \quad (1)$$

and apply least-squares techniques. In South Africa and France (Matheron, 1967a), the preference seems to be for

$$\textit{value at data point } = \textit{ value at a point on a random function} \qquad (2)$$

and application of moving averages. In fact, these two models are not really distinct.

It seems useful, without attempting to say what should be used, to try to explain the conceptual background of these methods, because this is not available in an elementary exposition. This area requires considerable statistical research; it has been mainly the preserve of theoretical statisticians. What little there is in the literature by way of applications has been written by geophysicists, so perhaps the earth sciences as a whole will help to create a new class of statistical methods (*see* some of the references in Whittle, 1963). It is for this reason that little is said about data analysis.

CONCEPTUAL PROBLEMS

Let me restrict the argument to the two-dimensional problem. Suppose we have a quantity y, measured at points (x_1, x_2) in a plane, distributed over some region R. Geometrically, we have points (with co-ordinates in three dimensions) (x_1, x_2, y). Through this cloud of points we want to put a smooth surface. We hope that this surface will show how y would vary with position (x_1, x_2) if there were no local or small-scale variation. The effect of large-scale disturbances is supposed to be incorporated into the smooth surface, which may be called the trend surface (Fig. 1). Thus,

$$\textit{observable surface } = \textit{ trend surface + variation surface.} \qquad (3)$$

The separation into two parts is made on the vague basis of large- and small-scale variation. Our interest may be in either part. The trend surface seems appropriate for interpolation and averages; the variation surface for the detection of anomalies, for example, discrete ore bodies. It is clear that unless we can gain more knowledge of the particular phenomenon under study, when we first observe it, or have a bright idea about the mechanism after a first look at the data, any analysis must be quite empirical. It follows that we cannot say any method is best.

Although I will discuss some specific mechanisms that might account realistically for some geological distributions, I will continue with a more general analysis of (3). In comparing (3) with (1), it can be seen that the two parts are assumed to be of a different nature: the trend is assumed to be a deterministic function and the variation to be random or stochastic. If (3) is compared with (2), there is no dissection but randomness appears again.

How might randomness enter? In the first place, y will usually be measured with error, that is, if the site is revisited, sampled, and the sample measured, a different value would be obtained. This is well understood and so can be ignored for the moment. Suppose now that our interest is only in the region R that was sampled at a finite number of points (I will later consider the case where region R itself has been chosen because it is typical of many similar areas); that is, our interest is in the population from which R was somehow chosen. There are infinitely many hypothetical surfaces passing near a finite number of points. Some will be more credible geologically than others. It is a standard tactic in science to deal statistically with an infinite number of hypothetical populations. Explicitly, we must define the class of admissible surfaces and suppose our particular case is a randomly chosen member of the class. Apart from measurement errors, then, it is in this way that randomness enters when we are concerned with only one area.

The population of admissible surfaces may be defined in several ways, corresponding to (1) or (2). I will return to this shortly.

If the region R has been chosen as typical of many other regions, even complete knowledge of the surface over R would be insufficient for conclusions about the other regions. It is necessary to define again a class of admissible surfaces and to regard the surface over R as a randomly drawn member of the class if we are to use statistical arguments.

To define a class of admissible surfaces, we are guided by scientific plausibility and mathematical and computational convenience. The older method is that

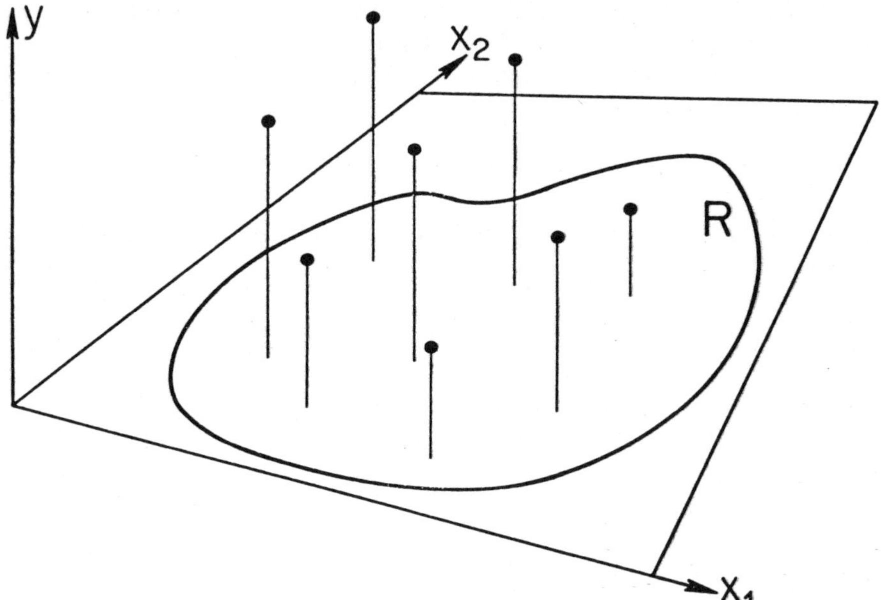

Figure 1. The value of the quantity measured at a point (x_1, x_2) is shown by the height of a vertical line above the point (x_1, x_2). Trend surfaces will pass nearly all the points, shown as dots.

of (1), and the commonest choices for the deterministic function are (a) an algebraic polynomial in x_1 and x_2 and (b) a trigonometric polynomial, that is, they are the first few terms of a two-dimensional Fourier series. It is usually implicit in papers written on this basis that the choice of function, the highest-degree terms used, and so forth, should make the random error have a zero mean, be uncorrelated, and have constant variance. For example, the most thorough study in the algebraic polynomial case (Grant, 1957) makes this quite explicit. Grant's method may be repeated using trigonometric polynomials, or any other set of functions. In detail, they depend strongly on having observations on a regular grid.

However, the model (1) method does not have this requirement and so has appealed to geologists, but the error should be studied if the model is to be checked at all with reality.

The errors referred to in the preceding paragraph should be thought of as the heights of a random surface $e(x_1, x_2)$ above the observing points even though we might have used other observing points. If $P' = (x'_1, x'_2)$ and $P'' = (x''_1, x''_2)$ are any two observing points and if we write $e(x_1, x_2) = e(P)$, the above assumptions read

$$E[e(P')] = 0, \quad E[e^2(P')] = \sigma^2 ,$$

$$E[e(P')e(P'')] = 0 . \tag{4}$$

Clearly the last assumption is unreasonable for many geological surfaces such as structural surfaces, thickness of stratigraphic units, and so forth, if P' and P'' are close together. It is reasonable if they are far apart, for this is the way spatial correlation is expected to behave, if one supposes that $e(P)$ describes small-scale disturbances only. It would be preferable to do an analysis in which the last two assumptions of (4) were replaced by

$$E[e(P')e(P'')] = c(P', P''), \tag{5}$$

where the covariance function $c(P', P'')$ decreased as P'' moved away from P'. If it is true that

$$c(P', P'') = c_1(P' - P''), \tag{6}$$

that is, that $c(P', P'')$ depends only on the vector $(x''_1 - x'_1, x''_2 - x'_2)$, then the error function is said to be a homogeneous random function. If it is assumed, however, that

$$c(P', P'') = c_2(r),$$
where
$$r = \sqrt{(x'_1 - x''_2)^2 + (x'_2 - x''_2)^2} , \tag{7}$$

the error function is an isotropic random function, because the covariance now depends on the length but not the direction of the vector separating P' and P''

In neither case does the average depend on the absolute position of P' and P''
These are the generalizations to two dimensions of the notion of stationarity,
now common in time series analysis which is the one-dimensional equivalent of
our problem.

The spectral methods of time series analysis are now familiar, particularly to
geologists who have followed the computer applications of the Geological Survey of Kansas group. The weakness of the model (1) work has come from ignor-
ing the two-dimensional time series aspect of the error model implied in model
(1).

Having come this far, one sees that it may be unnatural to separate the observable surface into a deterministic and a random function, because small- and large-
scale disturbance notions find a natural definition in spectral terms: high and
low frequencies. It may be better to define the class of admissible functions in
terms of averages throughout the class, to assume homogeneity or isotropy, and
to base the analysis on this assumption. I cannot, in this short paper, explain
how the analysis would then go, except to note that moving averages appear nat-
urally. It is not too different from the trigonometric equivalent of Grant's
(1957) method. Thus it is very awkward without regularly spaced data. Also
the weak assumptions generally made imply that one should have a considerable
number of data points. This is one form of model (2). I hope to deal with methods based on this model in another paper; meanwhile a list of relevant references
can be found at the end of Whittle's (1963) paper.

Matheron (1965) advocates another version of model (2). He does not assume that the averages are independent of the absolute position but that the averages of changes are. Explicitly, he considers random surfaces $f(x_1, x_2)$ such
that, for all (x_1, x_2), (h_1, h_2)

$$\tfrac{1}{2} E \left[f(x_1 + h_1, x_2 + h_2) - f(x_1, x_2) \right]^2 = \gamma(h),$$

where   (8)

$$h = \sqrt{h_1^2 + h_2^2} \quad .$$

This is not to imply that

$$s(x_1, x_2) = f(x_1, x_2) - f(0, 0) \qquad (9)$$

is an isotropic random function because (8) does not allow the calculation of
$c(P', P'')$ when P' and P'' are distinct. The extensive literature generated by
Matheron and his co-workers (Haas and others, 1967; Matheron, 1965, 1966a,
1966b, 1966c, 1967b, 1967c, 1968a, 1968b; Serra, 1967a, 1967b, 1968a,
1968b), particularly Serra, has only recently come to my attention. At the time
of writing, it is not clear how assumption (8), even with a particular choice for
the function $\gamma(h)$, is made to support a method of analysis. Matheron, however,
endorses the use of moving-average methods, which he attributes to Krige. In
fact, he uses "Kriging" to describe the act of taking moving averages. Matheron

writes that his methods do not require regularly spaced data points. More extensive accounts of Matheron's work will be found in Watson (1970, 1971, 1972).

To conclude this section, I return to measurement errors. If they are present, one must add to the random surfaces discussed above a random surface which is isotropic and in which the spatial correlation falls off very rapidly, that is, the equivalent to the white noise of time-series analysis.

SOME MECHANISTIC MODELS

There are two rather general ways of building probabilistic mechanisms that may have application to trend surface work, that is, that generate classes of random functions of geological pertinence. The key paper, with a large bibliography, is Whittle (1963). The paper by Matérn (1960) is also a basic source. This literature is all at a sophisticated mathematical level.

As an example of the first method: suppose ore bodies of different sizes are initially distributed randomly in the plane and that, after a given time, their influence is felt at neighboring points. For example, their contents may diffuse outward from their initial positions, which we might approximate by points. If the medium is homogeneous, $\xi(w, r)$ could stand for the concentration at a point, a vector r away from a body of mass w. If masses w_i fell at positions P_i, then a measurement at x, $y(x)$, free of measurement error, will yield

$$y(x) = \sum_i \xi(w_i, x - P_i) \qquad (10)$$

since for the ith mass, $r = x - P_i$. The sum in (10) is over all the masses. If the masses are distributed over the whole plane by a Poisson process so that the mean number in area A is λA, and if the probability of a mass lying in $(w, w + dw)$ is $f(w)\, dw$, independently of its position, the stochastic properties of (10) may be worked out easily. In particular, the process is homogeneous. If $\xi(w, r)$ depends only on the length of r, it is also isotropic. If it is relevant, then interesting problems arise because now the data should enable us to estimate λ and $f(w)$.

This model could also be used for linear reefs; one would then distribute lines rather than points. Another generalization would arise if the Poisson distribution is inadequate and has to be replaced.

The statistical treatment of data on these models is in its infancy. Models of type (10) are called moving-average representations. They have recently been used by Marcus (1970) to study cratering on the lunar surface.

Another class of models is suggested by classical mathematical physics, in which the mathematical description usually has the form

$$L\, y(x) = 0\,, \qquad x \text{ inside } R,$$
$$y(x) = e(x)\,, \qquad x \text{ on } S \qquad (11)$$

where S is the boundary of R and where L is a linear operator. For example, if

y = temperature at equilibrium, $L = \partial^2 / \partial x_1^2 + \partial^2 / \partial x_2^2$.   In usual practical cases, the temperature distribution over the boundary S determines the temperature in the interior R. One might expect $y(x)$ to be some average of the temperature $e(x)$ on the boundary, that is,

$$y(x) = \int_S b(x, x') \, e(x') \, dx' \; . \tag{12}$$

The averaging function $b(x, x')$ is related to the Green's function (*see*, for example, Courant and Hilbert, 1961) of the problem and is determined uniquely by L and S. Thus if $e(x')$ is a random function on S, $y(x)$ is a random in the interior. It will be seen that (12) is again a moving-average representation like (10). It is merely obtained in a different way. Fluid flow is described similarly, so it seems this approach should be relevant for some problems.

A much more difficult situation may be relevant to many geological problems; above I have written about boundary value problems in an isotropic medium with random boundary values. The converse situation—a medium whose properties vary randomly and with simple boundary conditions—is much harder to formulate and deal with.

ACKNOWLEDGMENTS

Research for this paper was supported by the Office of Naval Research under contract NONR 4010 (09) awarded to the Department of Statistics, The Johns Hopkins University.

REFERENCES CITED

Courant, R., and Hilbert, D., 1961, Methods of mathematical physics: New York, Wiley-Interscience.

Grant, F., 1957, A problem in the analysis of geophysical data: Geophysics, v. XXII—2, p. 309—344.

Haas, A., Matheron, G., and Serra, J., 1967, Morphologie mathematique et granulometrie en place: Annales Mines, v. XI, p. 735—753, and v. XII, p. 768—782.

Harbaugh, J. W., and Merriam, D. F., 1968, Computer applications in stratigraphic analysis: New York, John Wiley & Sons.

Marcus, A., 1970, Comparison of equilibrium size distributions for lunar craters: Jour. Geophys. Research.

Matérn, B., 1960, Spatial variation: Medd. Skogsforskyn. Inst., v. 49, no. 5, 144 p.

Matheron, G., 1965, Les variables régionalisée et leur estimation: Paris, Masson, 306 p.

—— 1966a, Structure et composition des perméabilites: Rev. de l'Institut Français du Pétrole, v. XXI—4, p. 564—582.

Matheron, G., 1966b, Genèse et signification energétique de la loi de Darcy: Rev.
de l'Institut Français du Pétrole, v. XXI—11, p. 1697—1706.
—— 1966c, Comparaison entre les échantillonnages à poids constant et à effec-
tifs constants: Rev. de l'Industrie Minérale, v. VIII, p. 609—621.
—— 1967a, Kriging or polynomial interpolation procedures: Canadian Mining
and Metall. Bull., v. LXX, p. 240—244.
—— 1967b, Eléments pour une théorie des milieux poreux: Paris, Masson,
106 p.
—— 1967c, Composition des perméabilités en milieu poreux hétérogène:
Methode de Schwydler et règles de pondération: Rev. de l'Institut Français du Pétrole, v. XXII—3, p. 443—466.
—— 1968a, Osnovy Prikladnoï Geostatistiki: Moscow, Editions Mir, 408 p.
—— 1968b, Composition des perméabilites en milieu poreux hétérogène: Crit-
ique de la règle de pondération géométrique: Rev. de l'Institut Français du
Pétrole, v. XXIII—2, p. 201—218.
Merriam, D. F., and Cocke, N. C., 1967, Computer applications in the earth
sciences: Colloquium on trend analysis: Kansas Geol. Survey Computer
Contr. 12.
Serra, J., 1967a, Échantillonnage et estimation locale des phénomènes miniers de
transition: (thèse), Nancy, 670 p.
—— 1967b, But et réalisation de l'analyseur de textures: Rev. de l'Industrie
Minérale, v. 49, no. 9, 14 p.
—— 1968a, Les structures gigognes: Morphologie mathématique et interpretation metallogenique: Mineralium Deposita, v. 3, p. 135—154.
—— 1968b, Morphologie mathématique et enèse des concrétions carbonatées
des minerais de fer de Lorraine: Sedimentology, v. 10, p. 183—208.
Watson, G. S., 1970, Quantification of geological variables: Johns Hopkins
Univ. Tech. Rept. 142.
—— 1971, Trend surface analysis: Mathematical Geology, v. 3, no. 3, p. 215—
226.
—— 1972, Prediction and efficiency of least squares: Biometrika, v. 59, no. 1,
p. 1—8.
Whittle, P., 1963, Stochastic processes in several dimensions: Internat. Statistical Inst. Bull., 34th sess., p. 974—985.

TECHNICAL REPORT NO. 124, DEPARTMENT OF STATISTICS, THE JOHNS HOPKINS
UNIVERSITY

MANUSCRIPT MODIFIED FROM TALK GIVEN AT QUANTITATIVE GEOLOGY SYM-
POSIUM AT THE GEOLOGICAL SOCIETY OF AMERICA ANNUAL MEETING IN
ATLANTIC CITY, NOVEMBER 10, 1969

THE GEOLOGICAL SOCIETY OF AMERICA, INC.
SPECIAL PAPER 146, © 1972

Systems Analysis:
Multidisciplinary Ecosystem Model

Richard A. Park
Department of Geology, Rensselaer Polytechnic Institute
Troy, New York 12181

and

John W. Wilkinson
School of Management, Rensselaer Polytechnic Institute
Troy, New York 12181

ABSTRACT

Systems analysis is the quantitative study of empirical and functional relations of a complex of interacting entities that comprise a specified system. Complex ecosystems can be properly understood only by utilizing the techniques and rationale of systems analysis. The multidisciplinary study of the Lake George, New York, ecosystem is given as an example of a comprehensive study in which geology is an integral part. Systems analysis involves the formulations of models, which are simplifications of the "real world"; the Lake George study has used a progression of such models, representing different levels of abstraction.

Conceptual models have been used to represent the more important components of the ecosystem and the principal transfer paths among these components; a comprehensive conceptual model has proved useful as a graphical representation of the interrelations among individual research projects. Data from the Lake George sampling program and from concomitant laboratory studies have led to the development of mathematical models of biologic, physical, and chemical processes. These have been expressed in computer logic and used in simulation studies; they have been evaluated by comparing predicted values with data collected in the field. Those models, such as the phytoplankton and hydrologic models, that yield realistic results and seem to be valid representations of actual processes are being analyzed in detail to determine how sensitive they are to various ecosystem changes, including man-induced effects. By combining functional models into a comprehensive ecosystem model, we can begin to forecast future consequences of management decisions and can apply the knowledge gained from all the associated studies in developing guidelines for environmental quality.

ABSTRACT MODIFIED FROM TALK GIVEN AT QUANTITATIVE GEOLOGY SYMPO-
SIUM AT THE GEOLOGICAL SOCIETY OF AMERICA ANNUAL MEETING IN
ATLANTIC CITY, NOVEMBER 10, 1969

Optical Processing:
an Alternative to Digital Computing

John C. Davis
Geologic Research Section, Kansas Geological Survey
and
Department of Chemical and Petroleum Engineering
University of Kansas
Lawrence, Kansas 66044

and

Floyd W. Preston
Department of Chemical and Petroleum Engineering
University of Kansas
Lawrence, Kansas 66044

ABSTRACT

The digital computer is undeniably an asset in the examination of many classes of geologic problems. Unfortunately, it is ill suited for handling pictorial information which constitutes a large percentage of geologic data. For certain kinds of problems, the inherent physical properties of optical lenses can be used to perform analyses that are impractical using a digital approach. For example, in a current study of pore structure in reservoir rocks, the pore pattern of an area 24 x 24 millimeters on a thin section was digitized, yielding more than one million data points. Spectral analysis was used to determine the relative contributions of spatial frequencies to the total porosity, but even with the Fast Fourier Transform, a two-dimensional spectral analysis of a single thin section is very expensive even on a large computer. In contrast, a proper optical lens system will produce a Fourier transform and map the power spectrum on film in a few seconds. A digital approach is more expensive by three or four orders of magnitude. Optical-processing methods are especially well suited for study of radar imagery air photographs and gross fabric patterns, as well as microscopic textures in rocks.

INTRODUCTION

Geology traditionally has been an observational, descriptive, and interpretative science. Until the last few years, measurement of variables and quantitative interpretation of data have played minor roles in its development. The forces operating in geologic processes are difficult to isolate and measure, and relating measured variables to fundamental forces in the earth has proved almost impossible. Because of the difficulty of delineating meaningful variables, geology has developed into a subjective science in which experience and intuition play dominant roles in research. It is true that geologists have tried since the earliest days of the profession to define and analyze geologic problems quantitatively. Until recently these studies have remained isolated instances, widely regarded as monuments to individual perseverance. Many geologists looked upon these early efforts as extravagant expenditures of time and manpower, often with results of dubious significance.

Most developments of quantitative methodology in geology have come from sister sciences. Geochemists, geophysicists, and ground-water geologists have been ahead of other earth scientists in the application of numerical methods. These methods, for the most part, were direct adaptations of techniques used in chemistry, physics, and hydrology; however, most geologists remained unaffected. This has been attributed to resistance to the quantitative approach, but probably there are additional reasons for the lack of adoption of these techniques.

Computers were first applied to geologic problems in the late 1950s and early 1960s (Harbaugh and Merriam, 1968). The earliest applications were straightforward extensions of conventional statistical and engineering techniques. Since then, there has been a rapid spread and expansion of quantitative methodology throughout the science. The expansion has come primarily because computers enable researchers to handle vast quantities of data with great rapidity and to make calculations that previously were impractical. Only by use of the computer can such sophisticated techniques as factor analysis, canonical correlation, and cluster analysis be applied.

Almost every facet of geology currently is undergoing a quantitative revolution. The survey of the Council on Education in the Geological Sciences, American Geological Institute (Reeves and Delo, 1970), is an attempt to isolate deficiencies in geologic education, and provides an insight into what geologists believe will be their future needs. One analysis (their Table 7) lists those skills requiring the most additional emphasis in future education. Of the top ten subjects, half are concerned with computing or statistics. In fact, the subjects "computer programming" and "data storage and retrieval" rank first and second, respectively.

The tremendous emphasis being placed on computer methods in geology suggests that many earth scientists regard the computer as a panacea. This is presumably based on the vague hope that massive number-juggling and data-crunching will somehow wring truth from confusion. Unfortunately, the hope is probably an illusion. Aside from naïve expectations of the power of digital computing, inherent limitations restrict the usefulness of computers in many geologic problems. A fundamental source of much geologic data is contained in planar surfaces at various scales. A typical example is the surface of the earth itself, as expressed in aerial photographs, radar images taken from satellites, or conventional maps. Smaller scale features such as those on walls of quarries or small areas of outcrop involve spatial relations that can best be described and recorded photographically. Thin sections are examples of data collections that are essentially two-dimensional. At the extreme end of the spatial scale, knowledge of ultramicroscopic features is derived from electron photomicrographs. All these various sources of geologic information have one common characteristic: they are forms of visual images. A photograph or similar presentation is by far the most economical and efficient way of recording and transmitting this information from one person to another.

OPTICAL VERSUS DIGITAL PROCESSING

The cliché that one photograph is worth a thousand words is a gross understatement. It is theoretically possible to pack a million bits of information into a square millimeter of film (Lansraux, 1965). With optical devices presently available it is quite practical to record and recover items with spacings of less than 10 micrometers (Reich and Dorion, 1965). It is evident that pictorial representations can contain tremendous quantities of information. Packing these data into a digital computer requires that three items must be stored for every sample point: the optical density or other measure of information concerning the point, an x-coordinate and a y-coordinate. If the image is sampled on a regular grid, explicit statements of x and y position are not necessary; rather, the location of data points can be inferred from their order in the array. Even so, the potential amount of data contained on one simple-appearing photograph is so great that it can overwhelm the capability of almost any existing computer. Figure 1 is a photomicrograph of a sample of the Muddy Sandstone (Cretaceous) from eastern Wyoming. Although this is essentially a monomineralic rock with a simple texture, it is nevertheless an extremely complex system in detail. The image can be simplified by distinguishing only between pores and grains. The same thin section dichotomized so that pores appear as black areas and grains as white areas is shown in Figure 2. All variables in the thin section have thus been reduced to a single compositional variable so that relations between grains and pores can be determined. To examine the microstructure of the pore network, the image must have a resolution sufficiently high to distinguish details of adjacent grains. That is, the image must be discretely sampled at intervals small enough to permit determination of sizes, shapes, and separations between grains. Several devices exist that sample and digitize images. One of the more sophisticated is an optical scanner directed by a small general-purpose computer. Figure 3 shows a computer listing of the results of digitizing a 24 x 24 millimeters segment of the thin section illustrated in Figures 1 and 2 at an equivalent optical resolution of 40 lines per millimeter. The digitized record contains 1,024 x 1,024 data points, or more than one million samples, and costs almost $100 to produce on the optical digitizer. The digital record occupies about 300 feet of 800-bpi (high density) magnetic tape. Just to list the digitized image on a high speed printer costs about

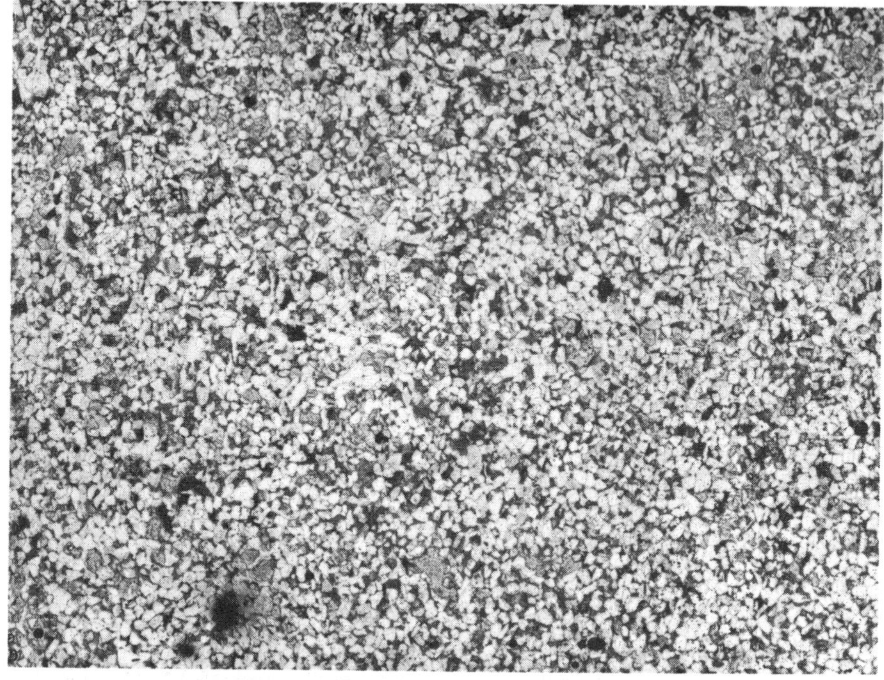

Figure 1. Photomicrograph of thin section of Muddy Sandstone (Cretaceous) from the Gas Draw Field, Wyoming. Pores in the predominantly quartzose sandstone have been filled with red plastic. Bar represents 1 mm.

$100. Computation of elementary statistics based on simple counts of lengths of grain intercepts would cost about $100 more. Analysis of spatial frequencies in the image by Fast Fourier Transform, with a program written in a general purpose language such as FORTRAN, requires about 30K of storage in a computer's central processor, four tape drives, and a drum or disk for four to six hours, costing several thousand dollars per thin section. These figures are quoted merely to emphasize the tremendous density of information packed into optical images and to stress that digital computing might not be the best way to analyze pictorial data.

For certain classes of geologic problems there are devices and techniques other than digital computers that enable researchers to perform quantitative investigations at great savings of time, effort, and money. These devices are particularly applicable in mathematical operations including integration and differentiation. Digital computers can integrate only by approximate methods involving successive summations, and therefore must trade running time for increased accuracy. Analog devices, including the electro-optical systems discussed here, perform these procedures directly. Resolution is limited by optical or electrical considerations, but analyses are performed almost instantaneously.

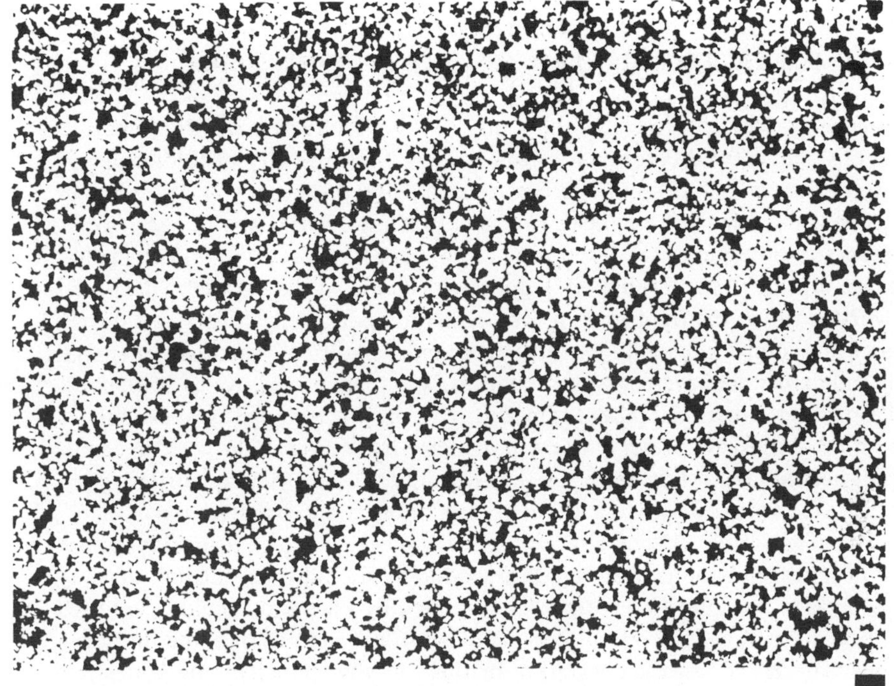

Figure 2. High-contrast photomicrograph prepared using Kodalith orthochromatic process film, which is insensitive to red light transmitted through plastic-filled pores. Image represents Figure 1, dichotomized into pore and particle states. Bar represents 1 mm.

Figure 3. Digitized representation of a segment of Figure 2. The image consists of a 1,024 x 1,024-grid of points measured by a flying-spot scanner and listed on the line printer of a computer.

POWER SPECTRUM PROCESSING OF POROUS MEDIA

A current research project of the Kansas Geological Survey and the Department of Chemical and Petroleum Engineering at the University of Kansas may illustrate some possibilities of one type of nondigital processor. This research is partly sponsored by the American Petroleum Institute and is one of a series of projects to study numerical descriptors of pore-grain geometry in porous media. Because flow of fluids through rock directly depends on the nature of the pore network in which the fluids rest, it is extremely important to understand the details of this complex system. Unfortunately most traditional methods of analysis yield only gross measurements of properties, which are insensitive to small-scale variations and directional inhomogeneity. This project is designed to examine details of the pore structure of clastic sedimentary rocks, and to do so in an economically feasible way. Furthermore, the techniques should yield quantitative results that are sensitive to size and orientation of pores and particles.

After considering several alternatives, it was decided to characterize the pattern of pores and grains seen in a thin section by the power spectrum of the pattern. This possibility was originally suggested by Fara and Scheidegger (1961), who limited their consideration to a one-dimensional analysis of a line through a porous material. The basic concept of mathematically characterizing pore structure by power spectrum calculations was subsequently expanded by Preston and others (1966). A computational approach to power spectrum characterization has been continued in the American Petroleum Institute project (Preston and Green, 1969; Preston and others, 1970).

If the image of a rock is dichotomized into pore and particle states, it can be regarded as a two-dimensional square-wave signal that can be analyzed by techniques developed in electrical engineering and time-series statistics. The two-dimensional square-wave nature of the pattern is emphasized in Figure 4, produced by electronically offsetting successive scans across the photograph. Particles appear as flat-topped plateaus and pores as flat valleys. One possible approach to analyzing such patterns is to determine contributions of different spacings between particles (spatial wavelengths) to the variance of the total image. This is analogous to frequency analysis of a time-variant electrical signal, except that attention is devoted to variance in space rather than variance in time. The term power spectrum comes from electrical engineering, where frequency analysis of an oscillating electrical current is used as a measure of its ability to do work.

In one dimension, the pore pattern of a rock can be represented by a signal such as Figure 5, taken from a traverse across a photograph of a thin section. This signal can be compared to a ray of white light, composed of many wavelengths. If a light beam is passed through a prism (Fig. 6), it is resolved into its constituent wavelengths. The prism is a frequency analyzer and the rainbow display is a transform of the beam from the time domain to the frequency domain. The relative abundance of different wavelengths in the beam is shown by the intensity of each color in the spectrum. The output of the prism is the power spec-

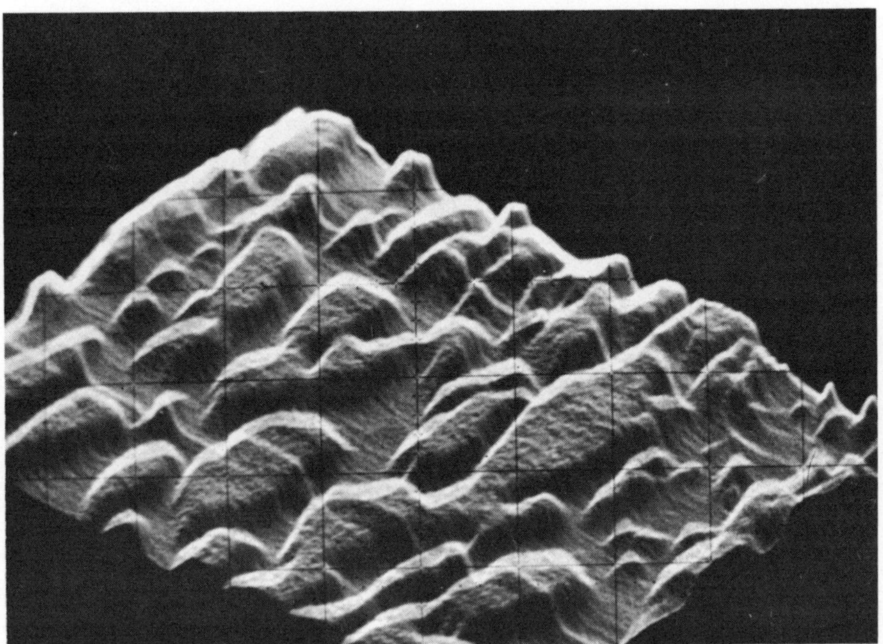

Figure 4. Dichotomized photograph can be considered as a two-dimensional square-wave signal, such as this display generated by the IDECS processor. Particles appear as plateaus and pores as valleys. This photograph is of a segment of Figure 2.

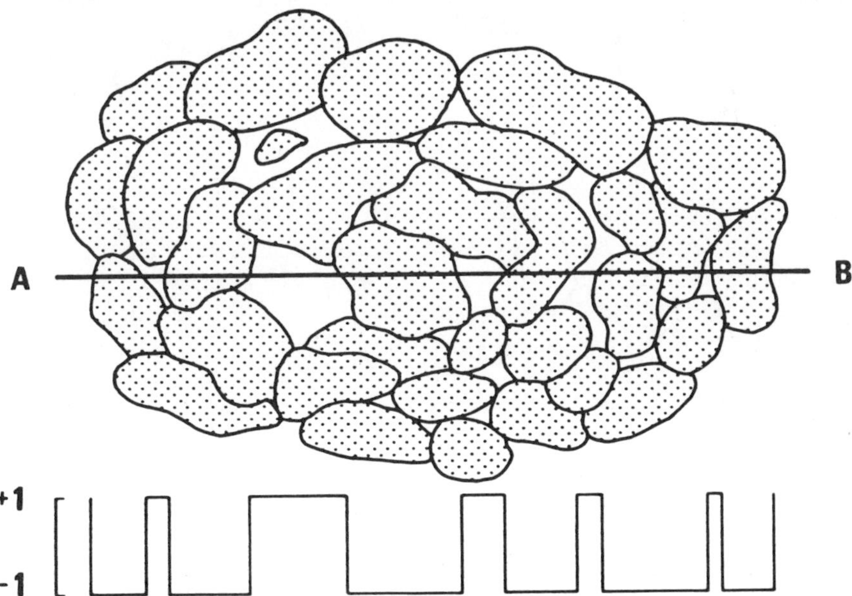

Figure 5. Traverse across a dichotomized thin section, represented as a spatial square-wave signal. Stippled areas represent grains. Signal has the value of +1 in pores, −1 in grains (Preston and others, 1970, Fig. 5).

trum of the light. The waveform in a ray of light changes with time, but the signal in Figure 5 changes with distance along the traverse. Nevertheless, a spatially changing signal can be analyzed in the same manner as a light ray, if a suitable analog to the glass prism can be found.

With certain simplifying assumptions, the normalized power spectrum of a one-dimensional spatial signal can be found from a Fourier transformation:

$$F(\omega) = \left| \int f(x) \, exp^{-j\omega x} \, dx \right|^2 \tag{1}$$

In equation (1), $f(x)$ is the input signal in the form of a continuous traverse, $F(\omega)$ is the normalized power spectrum, ω is spatial frequency, and j is $\sqrt{-1}$. Using similar conventions, the power spectrum of a two-dimensional spatial signal can be expressed by:

$$F(\omega_x, \omega_y) = \left| \int \int f(x, y) \, exp^{-j(\omega_x x + \omega_y y)} \, dx \, dy \right|^2 \tag{2}$$

Here, the input signal $f(x, y)$ is a continuous two-dimensional image and ω_x and ω_y are spatial frequencies in the x and y directions. If the input signal is discrete, these equations may be solved by finite Fourier methods, such as those described by Cooley and others (1967). However, if the discrete series has been generated by sampling an originally continuous signal, we encounter the problem of determining a sampling interval that will give adequate resolution, cover a sufficiently large area of the image, but which can be processed at reasonable cost. The number of operations that must be performed to compute the Fourier transform is

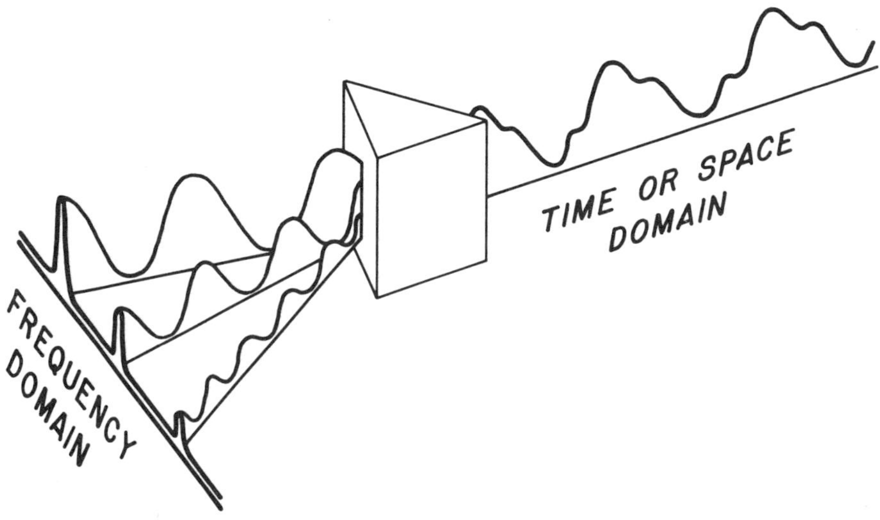

Figure 6. A complex waveform can be separated by an appropriate device or mathematical algorithm into its constituent frequencies. With polychromatic light, a prism serves as a frequency analyzer (Davis, 1970, Fig. 5).

proportional to $N \log N$, where N is the number of data points. This leads to extremely expensive computer operations, but even this is a significant improvement over conventional digital Fourier techniques whose running times increase at the rate N^2 (Cooley and Tukey, 1965).

Fortunately, it is not necessary to rely on digital computation to create Fourier transforms of images. One property of any simple lens capable of forming a real image is that it performs a Fourier transformation on the input signal. The geometric arrangement of lenses necessary to produce exact Fourier transforms is shown in Figure 7.

The ability of lens systems to produce Fourier transforms of images is well known; the essential mathematical relations were published by Michelson and Lord Rayleigh in 1892. Crystallographers have used optical methods to study crystal structures of minerals for many years. Their work has been summarized by Taylor and Lipson (1964). Barber (1949), in his study of the sea surface, seems to have been the first to analyze an image of a complex geologic subject by its optical Fourier transform. The procedures of optical processing were formalized by Cutrona and others in 1960 and were soon applied to the analysis of seismic records (Jackson, 1965; Dobrin and others, 1965). Pincus has published several articles on results of optically processing and filtering geologic images of various types (Pincus, 1966; Pincus and Ali, 1968), including an overview of the potential of optical processing in geology (Pincus and Dobrin, 1966).   Dobrin (1968) also has published an assessment of the potential of optical processing in the earth sciences. Recently, Pincus (1969a, 1969b, 1969c) has published a series of articles on the qualitative interpretation of power spectra of material having different grain-size distributions. Smith (1968) considered optical processors in a review of the use of automatic image analyzers for fabric studies. Some of

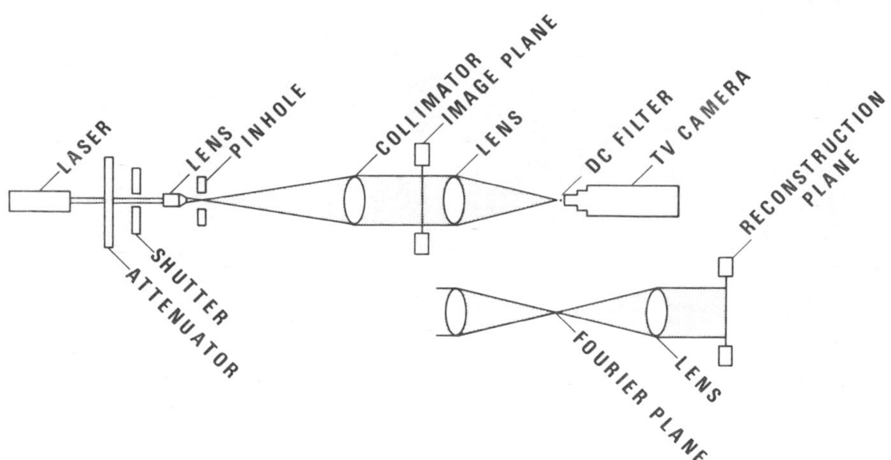

Figure 7. With the configuration of optical elements shown, an exact Fourier transform of the object will be produced at the transform plane. The object is in the front focal (image) plane and the transform is observed in the back focal (Fourier) plane of the transforming lens (Preston and others, 1970, Fig. 7).

the problems of optically processing images of microporous fabrics have been discussed by Davis (1970).

The complete Fourier transform is a function consisting of an amplitude spectrum and a phase spectrum. The phase is extremely difficult to record, but amplitude of light is expressed as its intensity, which is the square of the amplitude. It is necessary only to record the intensity of light across the focal plane in a linear manner. This may be done electronically with sensitive recording photometers or may be done photographically to create a permanent image that can be read on microdensitometers. Many commercial optical benches and other devices can be used to produce power spectra of two-dimensional images. The experimental system used by the Kansas Geological Survey (Fig. 8) consists of a 22-foot optical rail powered by a 1-milliwatt helium-neon laser; Fourier transforms are focused on the camera of a high-resolution closed-circuit television screen.

If the power spectrum is recorded linearly on photographic film, it can be isodensity contoured to give a quantitative power spectrum (Goodman, 1968). Frequencies present in the transform can be determined empirically using a ruled grating of known spacing. From the transform of the grating, a series of equally spaced concentric rings can be drawn, each corresponding to successive whole-number fractions of the fundamental spacing in the grating. Figure 9 is a typical calibration ring-set generated from a plate with 1-millimeter rulings. Successive rings represent spacings of 1 millimeter, 0.5 millimeter, 0.33 millimeter, and so forth. Using such calibration devices, the contribution of pore areas of specified size can be found. This is done by measuring the density of the recorded spec-

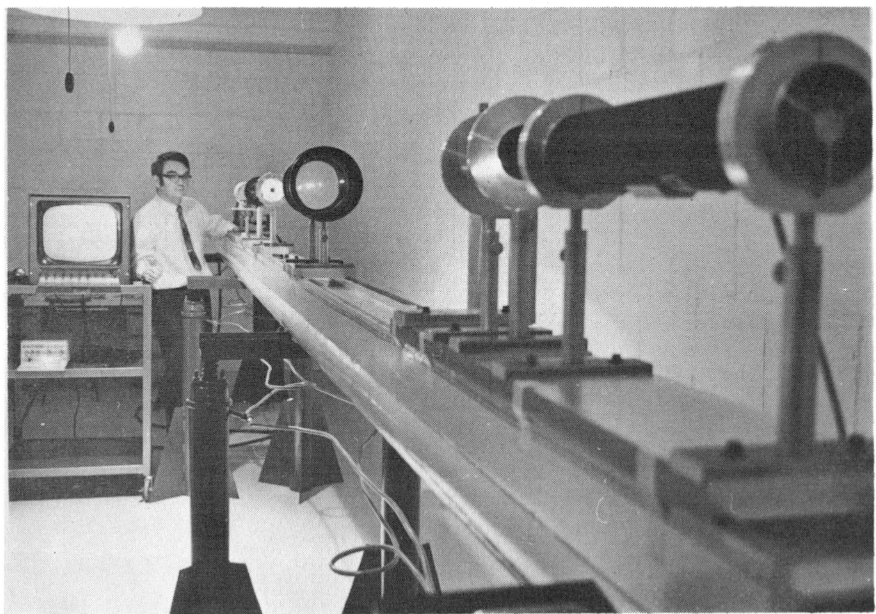

Figure 8. Optical processor built by the Kansas Geological Survey. The 22-foot long rail is suspended on a nitrogen gas cushion. Transforms are recorded on the television screen at left.

trum at positions on the transform corresponding to the desired spatial frequency. For example, the percentage of total pore area of the original image resulting from spacings between, say, 0.5 millimeter and 0.25 millimeter is proportional to the volume beneath the power spectrum between the calibration rings corresponding to ½ mm and ¼ mm. In other words, relative contributions to the total cross-sectional porosity, which result from different spacings across parts of pores, can be determined. It is easy to distinguish between transforms representing rocks having the same porosity but different pore size distributions. Furthermore, it is possible to quantitatively identify and measure the source of the differences.

Directional or vectorial properties are preserved in the Fourier transform. The contribution of spacings measured along traverses oriented between, for example, $10°$ and $15°$ across the original image is proportional to the volume beneath the

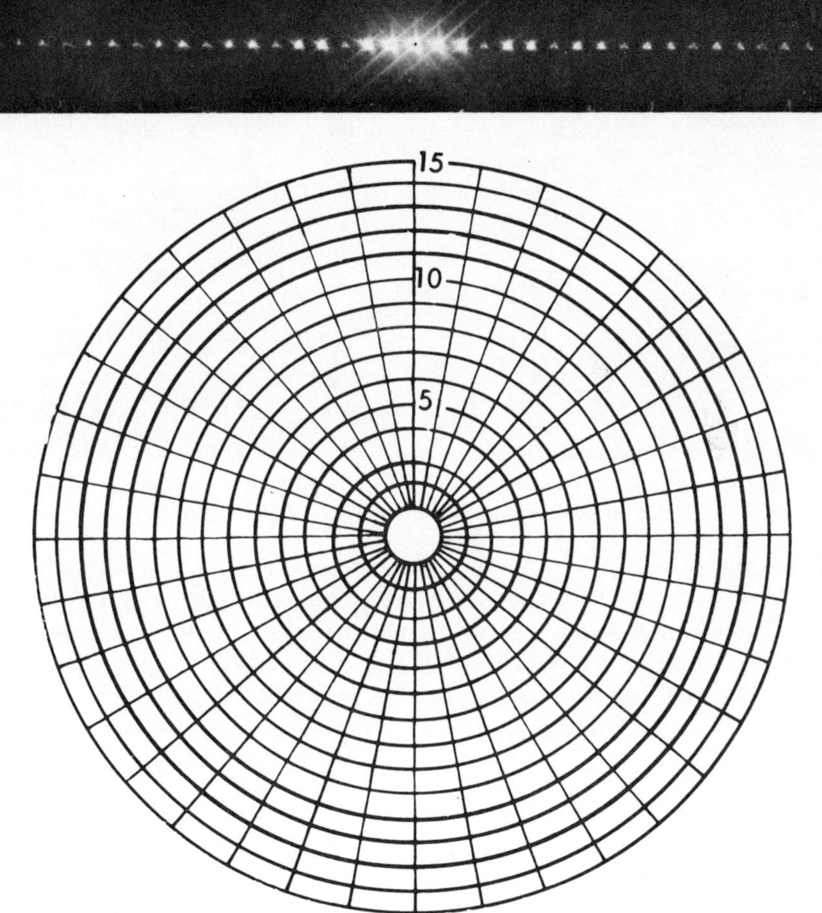

Figure 9. Spacings between points on the transform of a grid having 1-millimeter spacings (top) were used to construct the calibration-ring set. Each ring represents successive $\frac{1}{n}$ millimeter spacings on the transforms.

Figure 10. Transform of Figure 2 produced on the optical processor of the Kansas Geological Survey. This transform is typical of most porous sandstones studied.

wedge-shaped segment of the power spectrum bounded by radii at $100°$ and $105°$. This property allows quantitative assessment of directional inhomogeneities in the pore network. A single slice through the power spectrum along a specified radius is equivalent to the average one-dimensional spectrum of all possible traverses across the original image in a direction perpendicular to the slice. Therefore, within a single thin section, quantitative assessments can be made of directional inhomogeneities resulting from preferential orientations of pores or particles.

EXAMPLES FROM RESERVOIR ROCKS

Figure 10 shows a power spectrum of a porous sandstone; it was generated by the optical processor of the Kansas Geological Survey. In general, the transform appears as a round or elliptical area of gradually diminishing light intensity. Characterisitics of the optical power spectrum can best be illustrated with examples.

Each of the following illustrations consists of halves of power spectra from two samples chosen to show contrasting features. Parts of the original photographs used to construct the power spectra also are shown. In order to evaluate the transforms in a quantitative manner, it is necessary to measure relative brightness (intensity) of the spectrum, which is directly proportional to power. This may be accomplished in a variety of ways: direct measurement of light in the focal plane using a microphotometer, measurement of film density by microdensitometer, or use of television isodensity contourers. The IDECS processor, a digital video scanner (Dalke and Estes, 1968), was used to contour these power spectra at seven intensity levels. More sensitive devices are being investigated, bur for illustrative purposes the IDECS images are satisfactory.

Figure 11 is a comparison between Muddy Sandstone (Cretaceous) from eastern Wyoming, and Gaskell sand (lower Eocene) from California. Pore spacing in the Muddy Sandstone ranges from about 1 millimeter to the limit of resolution of the film. As a consequence of uniform grain size, distribution of pores is homogeneous and no obvious directional features are present. In contrast, the Gaskell is a poorly sorted, coarse-grained sandstone with more porosity concentrated in larger sizes. This is readily apparent in the contoured power spectra, where almost all power is concentrated in spacings greater than 0.2 millimeters. Median pore spacing of the Muddy Sandstone is less than half this size.

Contoured power spectra of replicate thin sections made from a core of St. Peter Sandstone (Ordovician) from Iowa are shown in Figure 12. The two spectra are essentially identical, demonstrating the expected consistency of the spectrum. Spectra of the two images are slightly elongate, indicating a shift toward shorter frequencies in a vertical direction. The thin sections were cut parallel to the axis of the core, so this may represent a subdued reflection of bedding.

Power spectra obtained from two samples of weakly consolidated Dexter sand of the Woodbine Formation (Cretaceous), collected on the Texas Gulf Coast, are shown in Figure 13. Sample A is a highly porous, well-sorted sand producing a power spectrum that drops off uniformly with decreasing pore size. Sample B also is unconsolidated, but in contrast is extremely poorly sorted with particle sizes varying from very coarse to very fine. The power spectrum therefore shows major variance contributions at long spacings. At pore spacings less than about 0.1 millimeter, the two sands are similar.

Contrasting directional properties are shown in two samples of Dakota Sandstone (Cretaceous) from Kansas (Fig. 14). One is a sandstone where microbedbing is recognizable through changes in grain size from one layer to another. Despite these changes, pore dimensions in one direction across the slide are not radically different from those in other directions. In contrast, microbedding is expressed primarily as changes in porosity in the second sample. These porosity differences result from selective cementation, producing elongate pores parallel to bedding. The heterogeneous pore structure produces an elongate pattern in the power spectrum.

This research project is still incomplete; yet to be established are relations between features of the power spectrum and more conventional measures of rock fabric and engineering properties. Samples must be examined by conventional

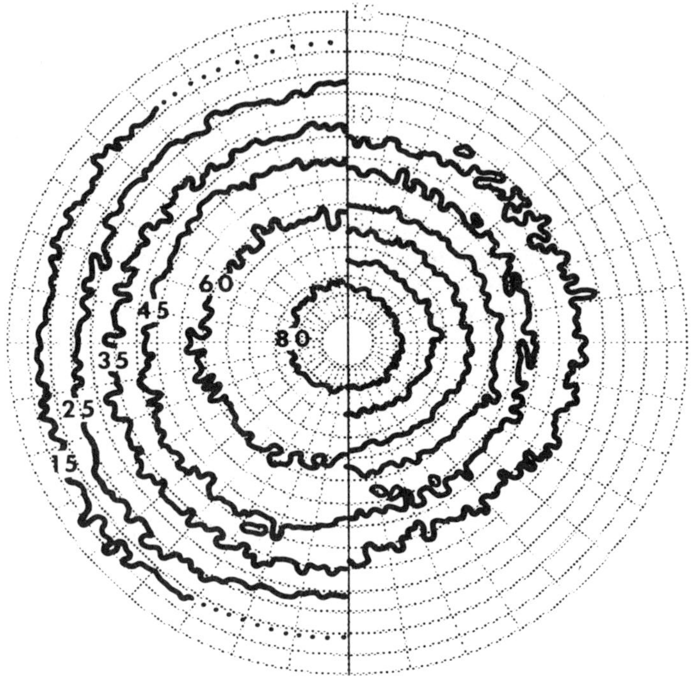

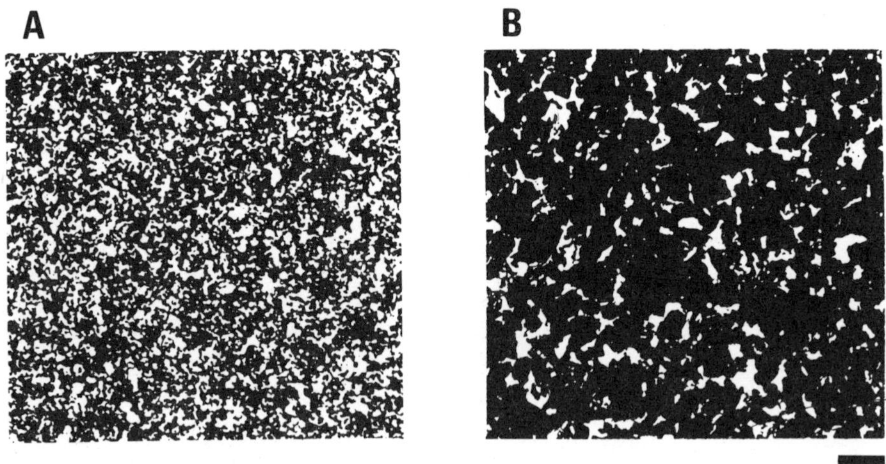

Figure 11. Contoured power spectra of Muddy Sandstone (A) and Gaskell sand (B). Contours determined by the IDECS processor and are relative percent power determined from optical density measurements. Calibration rings are $\frac{1}{n}$ millimeters. Bar on original images represents 1 mm.

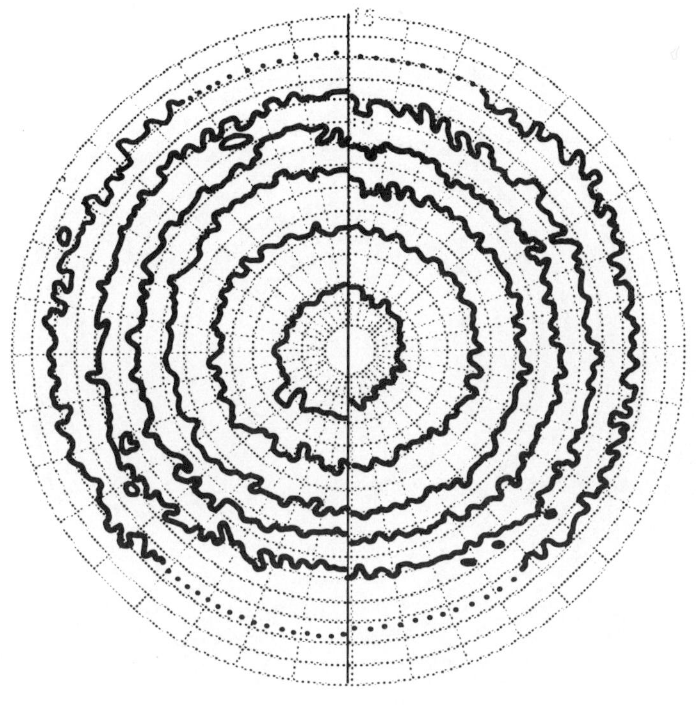

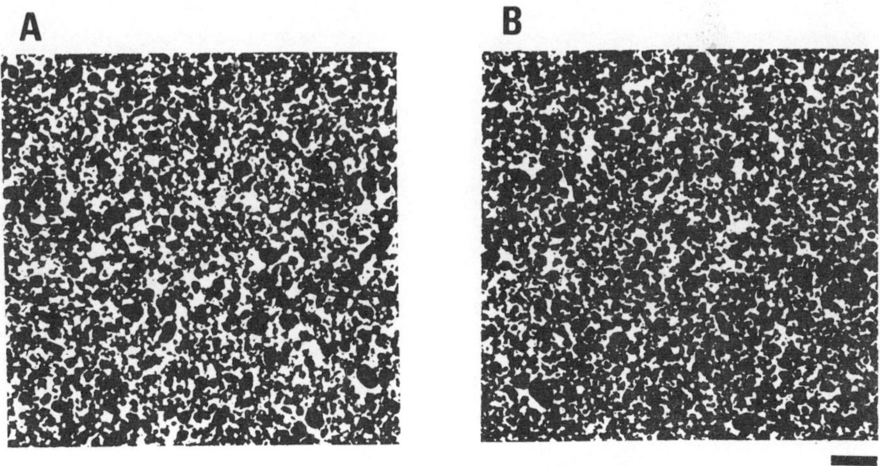

Figure 12. Contoured power spectra of two samples of St. Peter Sandstone (Ordovician) from Iowa. These demonstrate consistency of the power spectrum. Contouring and other conventions as in Figure 11.

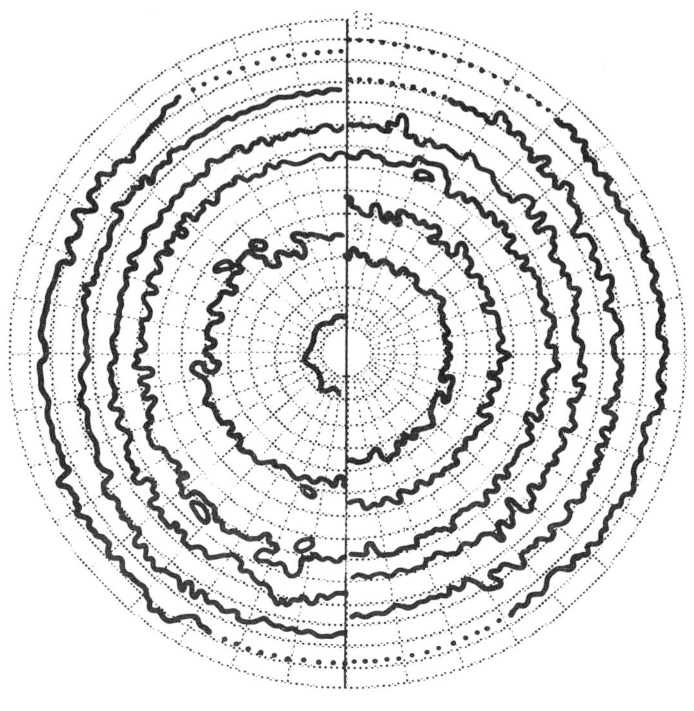

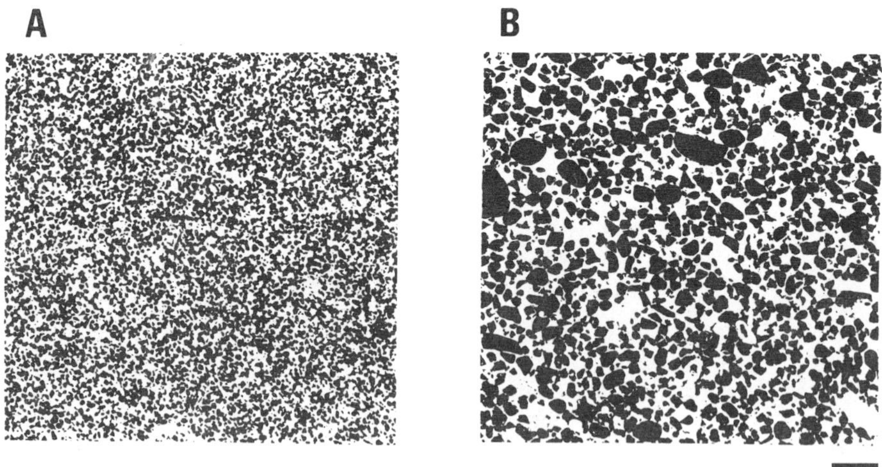

Figure 13. Contoured power spectra of two samples of unconsolidated sand (Woodbine Formation, Cretaceous) showing effect of sorting on pore distribution and power spectrum. Contouring and other conventions as in Figure 11.

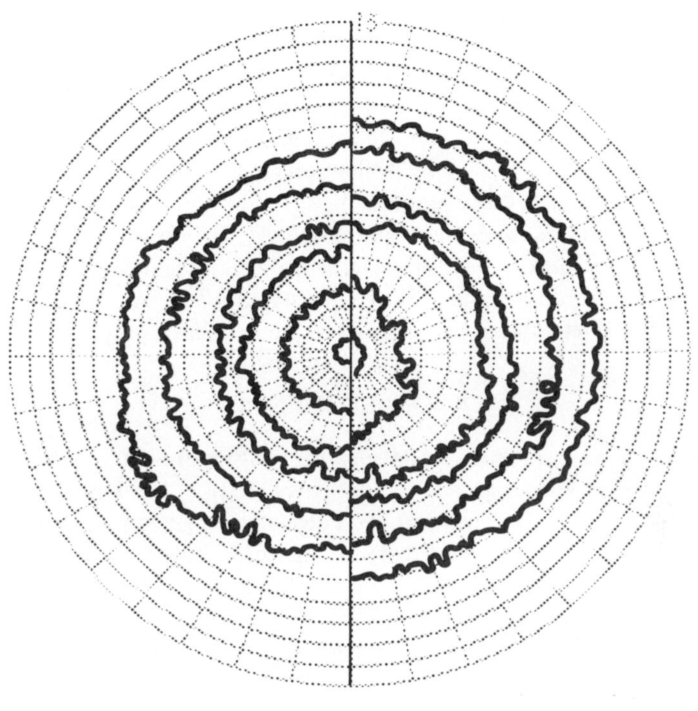

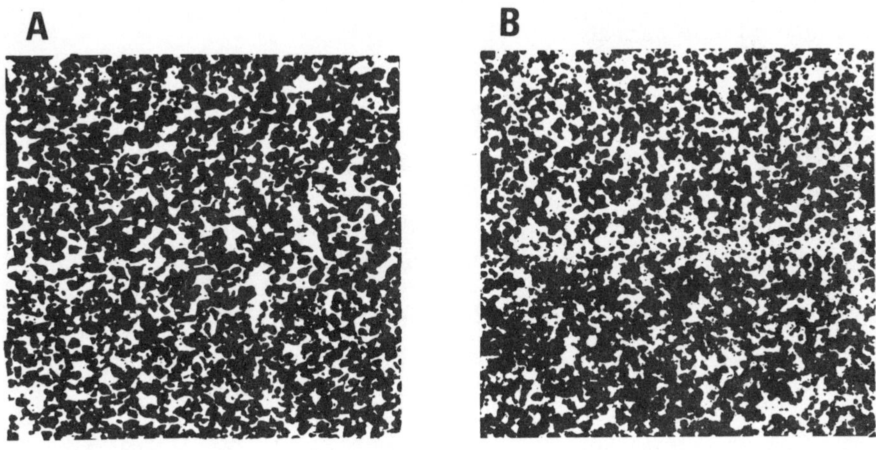

Figure 14. Contoured power spectra of two samples of Dakota Sandstone (Cretaceous), showing bedding effects on porosity. Microbedding produced by differences in grain sizes (A) does not create over-all inhomogeneities in the power spectrum. Microbedding due to differential cementation (B) produces a distorted spectrum at high power levels. Contouring and other conventions as in Figure 11.

techniques, including direct measurements of grains and pores on thin sections and determination of porosities and permeabilities. These variables must then be related to features of the optical transform. Until such correlation can be established, utility of the power spectrum technique cannot be fully assessed. The assessment must be based on a sample collection large enough that conclusions will be statistically valid. At this stage, the digital computer will be invaluable. Magnitude of the data set requires that automatic retrieval and analysis procedures be used. Thus, we will have come full circle, and a technique devised because of the inadequacies of digital computers will be integrated into a data analysis system in which both digital and analog processors complement each other.

ACKNOWLEDGMENTS

This report is part of American Petroleum Institute Project 103, Numerical Characterization of Porous Media, a cooperative study of the Kansas Geological Survey, the Department of Chemical and Petroleum Engineering at the University of Kansas, and the American Petroleum Institute. R. J. Sampson, Kansas Geological Survey, developed computer programs used. J. D. Farr, Pan American Research Corporation, advised on the construction of optical facilities. M. K. Wendland, Mobil Geophysical Laboratories, supervised digitization of images. G. M. Dalke supervised operation of the IDECS processor at the Center for Research, Incorporated, University of Kansas. N. S. Neidell of Seismic Computing Corporation kindly reviewed the manuscript and suggested appropriate modifications. These persons, their associates, and the companies they represent have contributed generously to the project, as have the many individuals who donated material for study and supplied analyses.

REFERENCES CITED

Barber, N. F., 1949, A diffraction analysis of a photograph of the sea: Nature, v. 164, p. 485.

Cooley, J. W., and Tukey, J. W., 1965, An algorithm for the machine calculation of complex Fourier series: Mathematics of Computation, v. 19, p. 297–301.

Cooley, J. W., Lewis, P.A.W., and Welch, P. D., 1967, The Fast Fourier Transform algorithm and its applications: IBM Research Paper, RC–1743, Yorktown Heights, New York, IBM Watson Research Center, 157 p.

Cutrona, L. J., Leith, E. N., Palermo, C. J., and Porcello, L. J., 1960, Optical data processing and filtering systems: IRE Trans. on Information Theory, v. 6, p. 386–400.

Dalke, G. W., and Estes, J. E., 1968, Multi-image correlation systems study for MGI, final report: Center for Research, Inc., Univ. Kansas, Contr. No. DAAKO2–67–C 0435, 245 p.

Davis, J. C., 1970, Optical processing of microporous fabrics, *in* Cutbill, J. L., ed., Data processing in biology and geology: Systematics Assoc. Spec. v. no. 3, London, Academic Press, p. 68–86.

Dobrin, M. B., 1968, Optical processing in the earth sciences: IEEE Spectrum, v. 5, p. 59–66.

Dobrin, M. B., Ingalls, A. L., and Long, J. A., 1965, Velocity and frequency filtering of seismic data using laser light: Geophysics, v. 30, p. 1144–1178.

Fara, H. D., and Scheidegger, A. E., 1961, Statistical geometry of porous media: Jour. Geophys. Research, v. 66, p. 3279–3284.

Goodman, J. W., 1968, Introduction to Fourier optics: New York, McGraw-Hill Book Co., 287 p.

Harbaugh, J. W., and Merriam, D. F., 1968, Computer applications in stratigraphic analysis: New York, John Wiley & Sons, 282 p.

Jackson, P. L., 1965, Analysis of variable-density seismograms by means of optical diffraction: Geophysics, v. 30, p. 5–23.

Lansraux, G., 1965, Contribution of diffraction optics to optical information technology, *in* Tippett, J. T., Berkowitz, D. A., Clapp, L. C., Koester, C. J., and Vanderburgh, A., Jr., eds., Optical and electro-optical information processing: Cambridge, Massachusetts Institute of Technology Press, 780 p.

Michelson, A. A., 1892, On the application of interference methods to spectroscopic measurements: Philos. Mag., v. 35, p. 280–299.

Pincus, H. J., 1966, Optical processing of vectorial rock fabric data: Internat. Soc. Rock Mechanics, Proc. 1st Internat. Cong., Lisbon, v. 2, p. 173–177.

—— 1969a, Sensitivity of optical data processing to changes in rock fabric, Pt. 1—Geometric patterns: Internat. Jour. Rock Mechanics and Mining Sci., v. 6, p. 259–268.

Pincus, H. J., 1969b, Sensitivity of optical data processing to changes in rock fabric, Pt. 2—Standardized grain patterns: Internat. Jour. Rock Mechanics and Mining Sci., v. 6, p. 269–272.

—— 1969c, Sensitivity of optical data processing to changes in rock fabric, Pt. 3—Rock fabrics: Internat. Jour. Rock Mechanics and Mining Sci., v. 6, p. 273–276.

Pincus, H. J., and Ali, S. A., 1968, Optical data processing of multispectral photographs of sedimentary structures: Jour. Sed. Petrology, v. 38, p. 457–461.

Pincus, H. J., and Dobrin, M. B., 1966, Geological applications of optical data processing: Jour. Geophys. Research, v. 71, p. 4861–4869.

Preston, F. W., and Green, D. W., 1969, Numerical characterization of reservoir pore rock structure, 2d Ann. Rept.: Am. Petroleum Inst., proj. 103, 84 p.

Preston, F. W., Green, D. W., and Aldenderfer, W. D., 1966, The use of statistical communication theory for characterization of porous media: Pennsylvania State Univ., Sixth Ann. Symposium, Computers in the Mineral Industries, p. B1–B20.

Preston, F. W., Green, D. W., and Davis, J. C., 1970, Numerical characterization of reservoir pore rock structure, Final Rept.: Am. Petroleum Inst., proj. 103, 98 p.

Lord Rayleigh, 1892, On the interference bands of approximately homogeneous light: Philos. Mag., v. 34, p. 407–411.

Reeves, R. G., and Delo, D. M., Jr., 1970, Requirements in the field of geology: CEGS Programs Pub. no. 5, Am. Geol. Inst., 38 p.

Reich, A., and Dorion, G. H., 1965, Photochromic, high-speed, large capacity, semirandom access memory, *in* Tippett, J. T., Berkowitz, D. A., Clapp, L.C., Koester, C. J., and Vanderburgh, A., Jr., eds., Optical and electro-optical information processing: Cambridge, Massachusetts Institute of Technology Press, 780 p.

Smith, A. R., 1968, Techniques for obtaining fabric data from coarse clastic sediments: Brigham Young Univ. Geology Studies, v. 15, p. 13–30.

Taylor, C. A., and Lipson, H., 1964, Optical transforms: Ithaca, New York, Cornell Univ. Press, 182 p.

MANUSCRIPT MODIFIED FROM TALK GIVEN AT QUANTITATIVE GEOLOGY SYMPOSIUM AT THE GEOLOGICAL SOCIETY OF AMERICA ANNUAL MEETING IN ATLANTIC CITY, NOVEMBER 10, 1969

The Geological Society of America, Inc.
Special Paper 146, © 1972

Systems Analysis and Model Building

Peter J. R. Buttner
Department of Geology
Rensselaer Polytechnic Institute
Troy, New York 12181

ABSTRACT

Systems analysis provides the procedural framework for the formulation, design, and analysis of symbolic models of systems. Usually the *structural* (organizational) aspects of a system are investigated by descriptive, analogue, or analytic modeling. Simulation modeling, which involves the merger of various structural models with a simulator and a simulation model, is usually employed to investigate the *functional* (operational) aspects of a system.

The sequence of procedures used in systems analysis forms a detailed statement of the execution of the scientific method. On using these procedures the model builder is required to examine his subject from new analytical vantage points. Operational research techniques such as linear programming, game theory, congestion analysis, and list processing can provide the model builder with various types of analytical support. In systems analysis the model builder also requires a high level of tactical support in such areas as data quality control, planning, allocation of funds, equipment, and time, scheduling, computer services, numerical analysis, and experimental design. Suitable methodology for both analytical and tactical support is generally available to most systems investigators from institutional computing centers.

As we continue to ask increasingly more complex questions concerning the structure and function of our modern and ancient environmental systems, we will find that the context of systems analysis is especially appropriate.

INTRODUCTION

This paper is a review of the systems-analytic approach to the formulation, design, and analysis of geologic models. Written principally for the novice model builder, it contains many examples of models that he can abstract for his own

use. An attempt has been made to illustrate the organization and diversity of the
systems-analytic approach. Details on models are limited to concept and structure; solution methods and manipulation techniques are available from a rich literature. The examples in this "roadmap" are taken from my dissertation, some
of which I have presented previously in courses, industrial seminars, and tutorials.

Formal systems-analytic techniques have become valuable components of the
modern environmental scientist's repertory of skills. He uses these techniques to
formulate questions, assemble answers, and to evaluate his performance. When I
began my work on river-basin simulation in 1960, much bootstrapping effort was
required to merge the systems-analytic concept with algorithmic procedures and
computer hardware. Initially this meant working with an IBM 650 and machine
language—FORTRAN was not yet available and data processing was quite different from what it is today. Equally important, systems procedures, computers,
and symbolic languages were unknown to all but a small group of geologists, and
effective communication between a model builder and his colleagues was sometimes difficult to achieve.

Today some introduction to systems-analytic procedures and automata is a
regular part of most graduate study. A large collection of pre-assembled or "can-
ned" programs, procedures, systems, and simulators is available to most students.
Computer manufacturers, institutional program libraries, and private individuals
can provide most of the necessary components to enable a student to begin a systems study. Several excellent texts (Griffiths, 1967; Krumbein and Graybill,
1965; Harbaugh and Bonham-Carter, 1970), written by geologists on the study
of various aspects of relevant geologic systems, are available to the student.
These sources illustrate the use of some of the important techniques of system
modeling—guidelines are proposed not only on what questions to ask in a system
study, but also on the formulation and context of those questions.

As is the case with many easily accessible tools and ideas, some inappropriate
applications have been attempted. However, the systems-analytic mixture of algorithms, computers, and proper geologic questions has produced some important
results. Because successful system modeling is a keen mixture of art and science,
which often only experience can bring together, it has taken some time for the
technique to be accepted as an important tool. When properly executed, systems
analysis is an elegant expression of the scientific method. In the systems-analytic
approach, the use of rigorous schedules and procedures is required to ensure adequate specification of the system's components, their organization and their function. Such formalization enables the systems scientist to describe complexly organized structures (statements, procedures, notations, and symbolic models) in
algorithmic terminology and in a context acceptable for electronic data processing. In many instances a computer is not always a necessary part of a system
study. However, because often we do not understand enough about a system to
properly describe its initial state, we are forced to assemble a first approximation,
set it in operation, and modify it by cut and try. This heuristic approach to system definition, which may require many iterations with extensive calculation and
data processing, is done best with a computer.

Named after an Arab mathematician, an algorithm is a sequence of operations that can be used to execute a general procedure. Algorithms are used to calculate such parameters as moment measures, structure factors, and chemical norms. Translation of algorithms into synthetic computer languages makes them available to machine processing and use in modeling. In the United States, FORTRAN (FORmula TRANslation) and its linguistic descendents are used by many systems scientists for model description. Sometimes, because of special aspects of the system under study, the model builder finds it an advantage to build his own synthetic language for modeling. My SILTS (System Interrogator and Lithosome Trend Simulator) is such a problem-oriented language (POL) (Buttner, 1965). POLs have been developed for water resources (MIDAS), statistics (BASIS), simulation (SIMSCRIPT), engineering (COGO), and many other activities. In the last two years my students and associates have used at least 72 different POLs in connection with our natural resource planning and management study of the Norman's Kill Basin (Buttner and Wachmann, 1972). All these synthetic languages follow the algorithmic format in their design and use. For instance, one synthetic language, ALGOL, is especially convenient to work with because of its concise procedural expressions and logical structures.

As the foundation of a computerized operation or program is its algorithm, it is much more useful to publish the algorithm in a journal or bulletin than to print a computer program. The program is language-dependent (and perhaps machine-dependent), while the algorithm is a general statement that can be translated into any appropriate synthetic language.

Increasingly, the systems-analytic finesse of the earth scientist is found to be a vital component in the study and analysis of environmental problems and a wide variety of social, economic, and political problems. As a consultant I have used these techniques in work related to criminal justice, social welfare, municipal government operations, forest management, natural resource planning, educational administration, ecosystematics, river-basin studies, utility power control systems, research planning and management, and various petroleum industry studies. Although all of the examples used in this paper deal with geologic systems, most can be used in many other areas. I hope that other applications will be apparent to the reader.

SYSTEMS

A system may be defined by a group of functionally related properties, processes, responses, algorithms, or models. Systems are usually organized into some order or pattern of the type: primitives—subdomains—domains. A corresponding sedimentologic set might be: sand grains—beds—lithosomes. The concept of a multi-input systems approach to problem definition and solution is not new to the earth scientist. What is different, however, are the requirements for extensive specification and detail necessary for the formal execution of systems analysis. For instance, the student of an ancient river basin must assemble his best estimates of various climatologic and hydrologic parameters if he wishes to simulate its patterns of stream ordering.

The systems-analytic approach requires that once we have delimited an entity of interest and committed ourselves to its study, we ask:

1. To what system does the entity belong?
2. What is the system's structure and function?
3. What are its process elements and what responses do they generate?
4. What is the system's simplest strategic framework?
5. How can I best model that framework?
6. What kinds and amounts of data do I need to build and test my model?
7. How do I evaluate the validity of my model?
8. What general applications does this model have to the characterization of the structure and function of other systems?

Systems analysis is the orderly, rigorous approach to answering these arguments in proper contexture and sequence. The general pattern of systems analysis is illustrated in Figure 1. This plan can be expanded into the sequence of procedures used to study and model a system. The key operations for the study of natural systems are shown in Figures 2 and 3 (expansions of Fig. 1).

Stage one of a system study involves 1 through 6 and stage two is concerned principally with 7 and 8. However, as shown in Figures 2 and 3, feedback mechanisms operate both between and within each stage, and the resolution of each argument is a dynamic procedure subject to continual revision during the study. It is not uncommon for the experienced analyst to model these feedback mechanisms and explore their effect on his basic formulation. In linear programming models this is called sensitivity analysis.

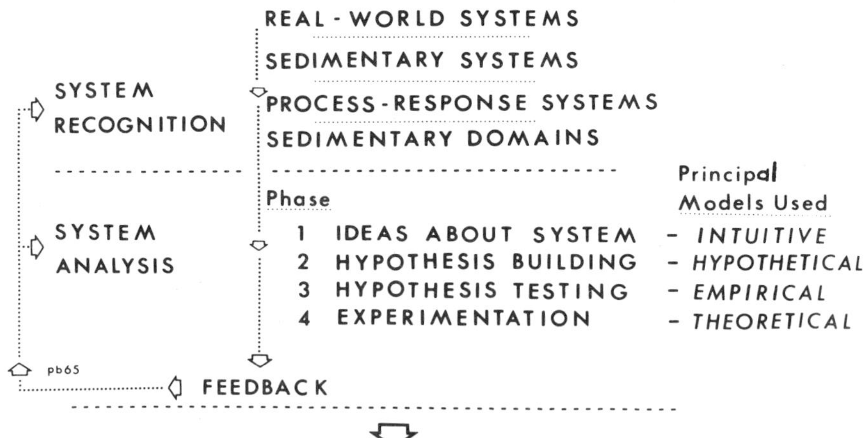

Figure 1. Systems and models. A system study involves system recognition and analysis. The main model types used in system analysis are identified as intuitive, hypothetical, empirical, and theoretical (Buttner, 1966a).

DATA MANAGEMENT

A model, no matter how elegant, is only as valid as the data used to generate its structure. Data management involves the collection, assembly, reduction, storage, quality control, and analysis of data elements. Some data management is used in all studies, but in the systems-analytic context, an organized data base of elements whose accuracy and precision are known is vital. A data processing plan

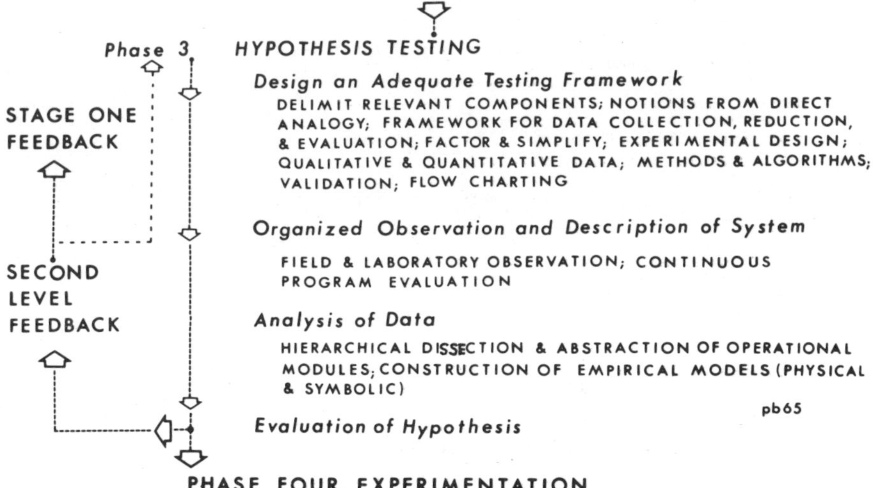

Figure 2. System analysis, stage 1. This stage includes the essential planning stages in the selection and definition of the system under study. Most geologic system studies require a field reconnaissance for the development of preliminary hypotheses. (Buttner, 1966a).

Figure 3. System analysis, stage 2. Four major activities are recognized: assembly of testing framework, field study, data analysis, and hypothesis testing (Buttner, 1966a).

that I have designed and used is shown in Figure 4. This plan consists of a set of functions that can be separated and operated independently. The system, in whole or parts, may be operated manually if required by the analyst. At this writing, it is being used with a large computing system to support a land-use and natural-resources-basin study, and with a small computing system, to monitor and model the flow-regime dynamics of a small river system. It is managed with a POL I invented for geologic studies. This plan can be used equally well with large or small computing systems.

A technique that may be used whenever scheduling is required—either in the execution of the study or in one of the study's models—is Program Evaluation and Review Technique (PERT). When PERT is used to model the research program, the researcher can get some idea of the schedule of critical events in his study and the time frame for each event. For instance, when a study requires stratigraphic field data, a stratigrapher can usually provide an estimate of his time, material, and support requirements. This information, together with similar data on other aspects of the study, is used to develop the study schedule in a PERT context. This PERT description is periodically solved and its computed schedule compared with

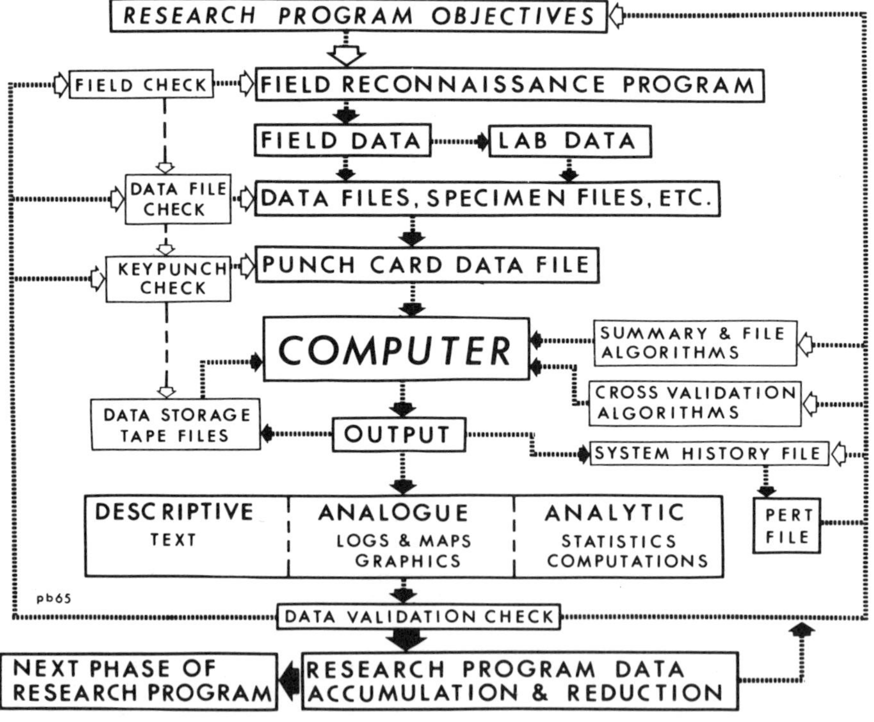

Figure 4. Data management support. A simple, effective algorithm is shown for the collection, assembly, review, and analysis of data. This plan is generally independent of computer type, programming language, and research objectives. The PERT option permits the researcher to monitor the schedule and performance of each aspect of his study—from field mapping to laboratory operations (Buttner, 1966a, 1968).

the real time project schedule. Activities that are behind schedule or improperly described in the PERT model are indicated, and appropriate actions by the analyst may be taken to either keep the project on schedule or to modify the project's objectives. This schedule can be updated as the study progresses and as additional data are developed. In those cases where PERT has been applied to new areas of investigation, it is often not possible at the onset to identify correctly all relevant aspects of the study, model, or system. In such cases the analyst can interact with his study and the PERT network to continually modify his objectives. For instance, PERT can be useful when the model builder wishes to keep concurrent sets of sequentially operating processes (for example, a set of developing offshore bars) keyed in time with a schedule of events (for example, maturation of a coastline). A simplified schematic of a PERT network is shown in Figure 5.

In a systems study proper data management requires an understanding of data quality control procedures. Figure 6 illustrates the need for some quality assurance in the data collection process. The researcher should not think that because he uses a computer his data are unmollested—key-punch operators, tape handlers, computer hardware, and some programs often insert their own "interpretations."

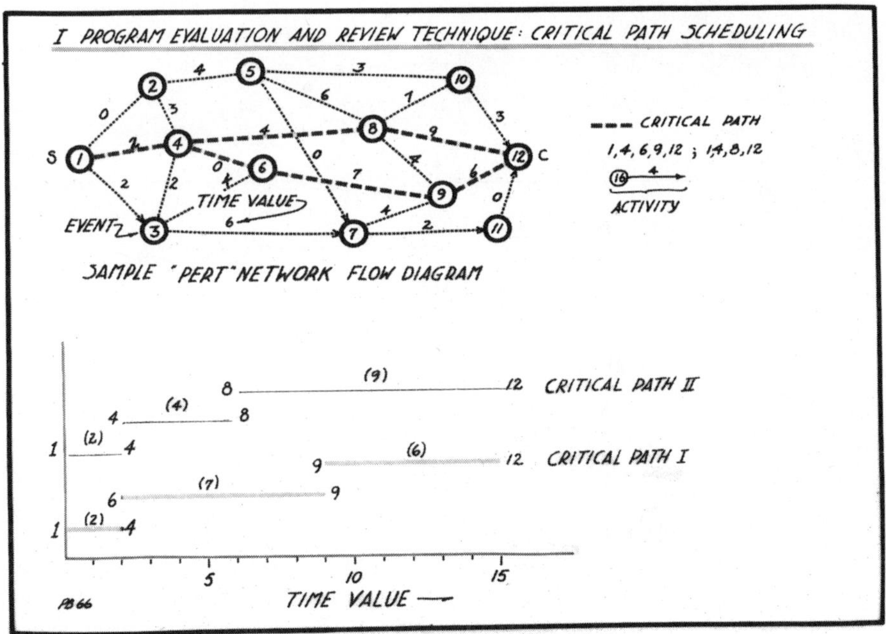

Figure 5. Program evaluation and review technique (PERT). Network is composed of numbered response events (circles) with process activities (arrows) connecting those response events that are sequentially dependent. Each activity is assigned a value for the time (or cost, or amount of rippled bedding, or percentage of gastropods) required for its "completion." This mathematical technique can be solved by paper and pencil, but usually the solution of all but the simplest networks requires a computer (or a team of analysts). PERT selects the critical pathway through the net. On this path are all those response events and process activities that must be executed before the network can be closed. This network is a greatly simplified characterization of a bed-form development plan for a small river basin (Buttner, 1967).

Clearly, data quality control is a most important aspect of experimental design and hypothesis testing. The earth scientist will do well to read Griffiths (1967, p. 317–320) and Duncan (1965) on data quality assurance.

Proper data management means that estimates of the accuracy and precision associated with each data element are available to the system analyst. He can expect that the data generators (observers, machines, instruments) of any dynamic system will exhibit spectral variations during measurement, comprising both inherent variations and those which are the result of the interaction of the generator and the measurement process. Consider, for example, the natural variation in gage height at one of the author's field stations caused by natural downstream traveling waves passing the station (Fig. 6, left). The errors associated with the sensing, transmission, and reporting of these data have been identified, and it is possible to predict the reliability of each operation using quality control procedures. When a group of operators monitored the discharge sensing equipment, at least four main types of "reporter-equipment" interactions were evident (Fig. 6, right). Biased reporting, out-of-control reporting, and their combination were all apparent. Attention to proper data management procedures is vital to successful systems analysis.

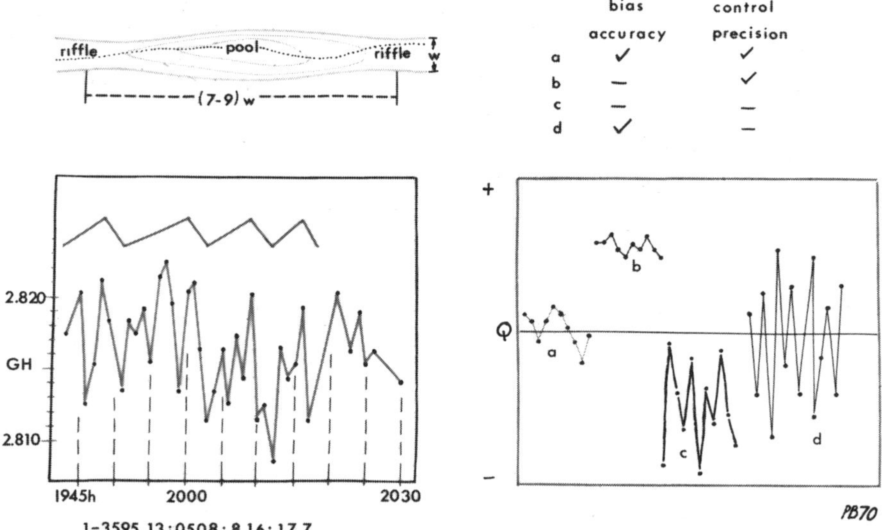

Figure 6. Field station quality control. At left is shown the variation in gage height (GH) measured in inches with time (1945-2030 GMT). This gage is in proximity to USGS station 1-3595.13, and these data were collected on 050871. The drainage area and discharge at 1945 GMT was 8.16 sq mi and 17.7 cfs, respectively. A flow variation model that follows a 2:1, rise-to-fall pattern is reflected in bed-form adjustment dynamics.

Twenty-six students in fluvial sedimentology made a series of discharge measurements with flow-metering gear in proximity to 1-3595.13; each student made from 10 to 13 repetitions. At right is a plot of four types of operator-measurement process interactions, (student operator types a, b, c, d) shown as deviations from the discharge time-trend, Q. Some operators, no matter what their intentions, education, sex, training, or profession, seem to interact with both measurement machines and procedures to generate data that are inaccurate and imprecise (b, c) (Buttner, 1966a, 1972; Buttner and Wachmann, 1972; Griffiths, 1967).

SYMBOLIC MODELS

Models are representative descriptions of some entity; symbolic models use symbols to perform the description (Buttner, 1966a). These symbols may be those of a natural language or a synthetic language devised for the purpose. Three types of symbolic models are used in system studies: descriptive, analogue, and analytic (Fig. 7). The development of a descriptive symbolic model is usually a necessary first step in the systems-analytic study. These models form the basis for subsequent design and assembly of analogue and analytic structures. Analogue models are more detailed extensions of the descriptive types, and in some studies the analyst proceeds directly from descriptive to analytic structure.

DESCRIPTIVE MODELS

Descriptive models are shown in Figures 8, 9, and 10; each model is organized in a different fashion. The model of the carbonate textural spectrum, Figure 8, is a single-sequence descriptive model. It represents an attempt to capture the essential textural contrasts that develop as carbonate sediment (in the upper left) passes through initial modification (right side, top to bottom) to dolomite and then on to "dedolomite" (left side, bottom to middle). The central view shows the complete spectrum of textures starting with sediment at the top and ending with "dedolomite" at bottom. This model was developed as part of a study of the relations between the various textural aspects of carbonate aggregates used in

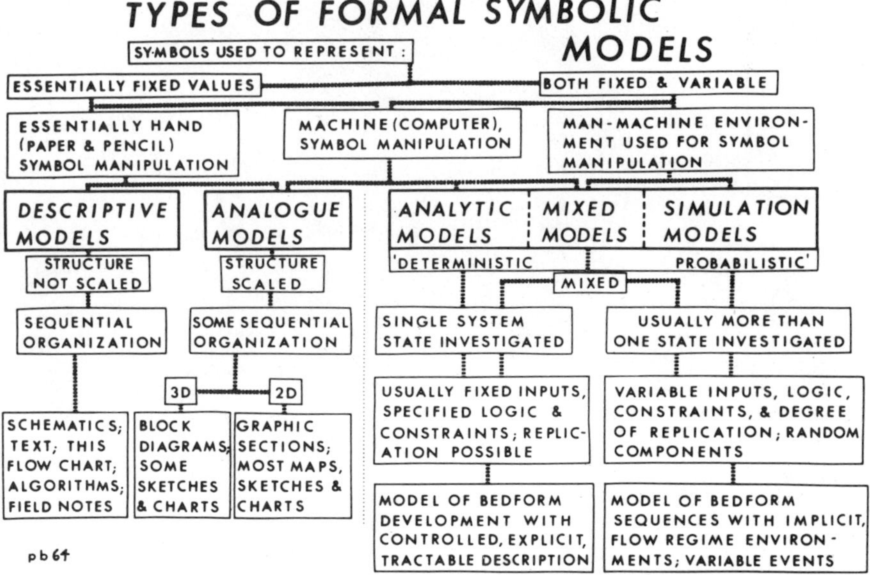

Figure 7. Types of formal symbolic models. Usually descriptive models are developed to define the work area. As the study progresses, information is assembled for the development of analytic and simulation models (Buttner, 1966a, 1966b, 1968).

Portland cement concrete and the chemical degradation of road surfaces. Later, based on this structure, it was possible to devise sets of analytic models for aggregate selection, concrete manufacture, and for evaluation of road surfaces.

It is possible to question the manner in which sediment is processed as it passes through various sedimentary domains. For instance, consider the migration of sand waves in several second-order streams of the same third-order system. Each second-order stream contributes sand to the higher order stream which must service this load according to some discipline or set of rules. The queueing format provides a modeling structure that can be used to develop descriptive models of competitive service systems. A descriptive queueing model and several physical-stratigraphic examples are shown in Figure 9.

The first example of a descriptive model was Figure 8, the algorithmic Carbonate Textural Spectrum, whose structure was that of a single sequence of events. Figure 9, the Descriptive Queueing Model, is also an algorithm consisting of a sequence of procedures, but is useful over a broad range of studies. The Fluvial Process-Response Model of Figure 10 is patterned in the same manner: as a sequence of process activities and response events. However, this model structure contains several major feedback loops that can modify the model's details during execution. Three common types of descriptive models can be recognized: sequential particular (Fig. 8); sequential general (Fig. 9); and sequential recursive (Fig. 10) (Buttner, 1966a).

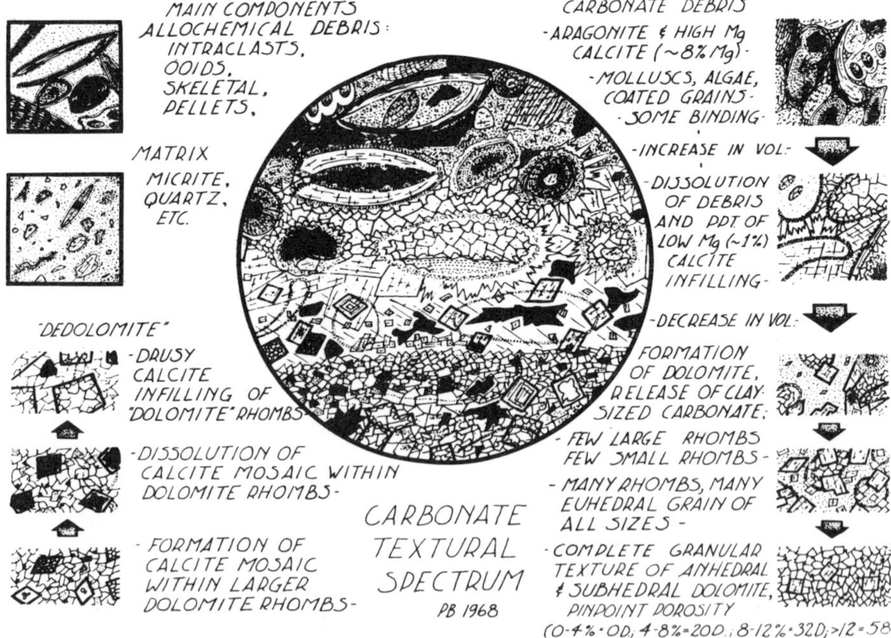

Figure 8. Carbonate textural spectrum. Textural variation of carbonate rock is shown from initial stages (upper left) through dolomite (lower left) to "dedolomite" (lower to middle, on right). Central sketch shows composite of textures with sediment at top and "dedolomite" at base. Scales are adjusted to emphasize features (Buttner, 1969b).

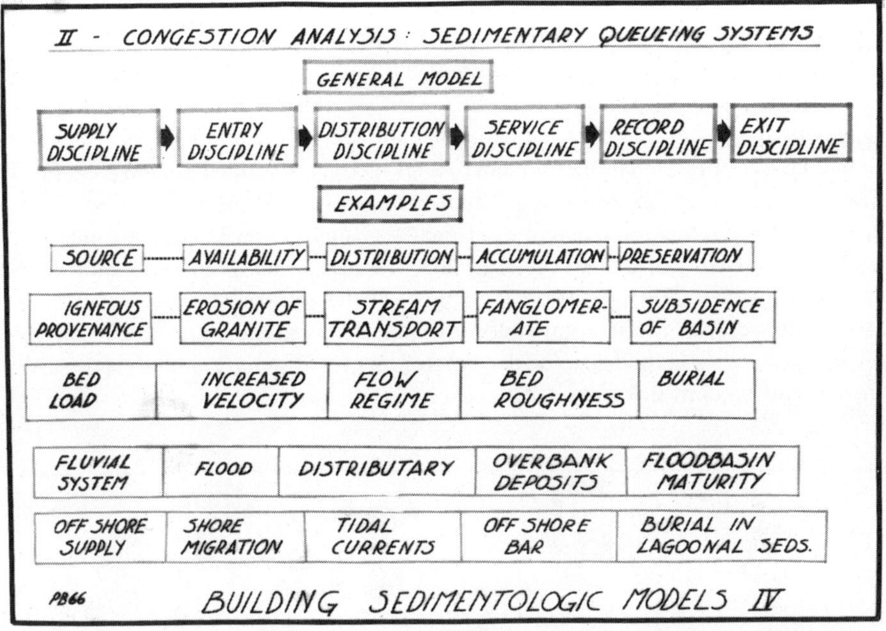

Figure 9. Descriptive queueing model. A queue is a collection of process activities and their corresponding response events that are waiting for service. Questions on basin evolution, landscape, morphology, bed-form modeling, shoreline development, and offshore bar migration have been cast into the queueing model format (Buttner, 1966a).

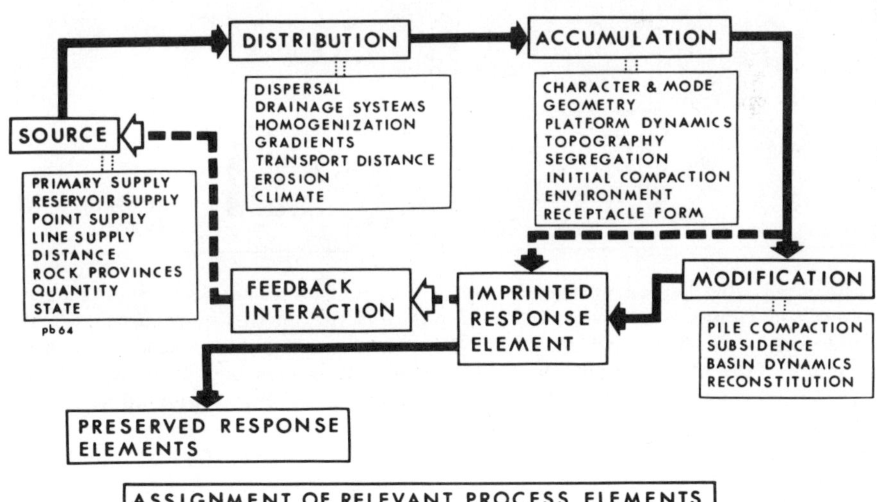

Figure 10. Fluvial process-response model. Four process activity domains are shown: (1) those related to the supply of available sediment and its state, (2) those related to the distribution system and its manufacturing aspects, (3) those related to the site of accumulation, and (4) those related to the regional physical-stratigraphic patterns. Major feedback loops are shown. This model provided the framework for an analytic-simulation model (Buttner, 1965, 1966a, 1966b).

ANALOGUE MODELS

An analogue model is a scaled, semi-quantified, and detailed representative of a descriptive model. The three-dimensional structure shown in Figure 11 is an analogue model of the rhythmic sequence developed by a braided stream system. In Figure 11, the map shows the extent of the outcrop pattern of Rhythm Four, a rhythmic continental sequence in the Genesee Group (lower Upper Devonian) of southeastern New York. This illustration represents a simplified expression of the details of Rhythm Four, yet it portrays most of the essential components necessary to characterize the structure and function of this subdomain. The assembly of this analogue was the basis for the development of simulation models to study ancient and modern fluvial systems.

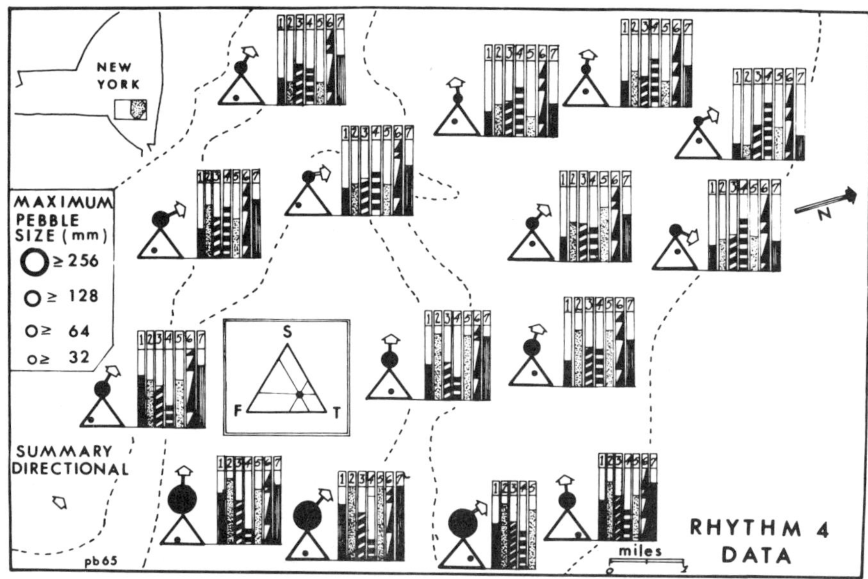

Figure 11. Rhythm four analogue model. The Upper Devonian of the Catskill Complex of New York consists of sequences deposited in glaciofluvial, fluvial, lacustrine, paludal, estuarine, coastal marsh, bay and lagoon, and various tidal environments. A generalized analogue profile of one of these sequences, a predominantly fluvial domain, is illustrated. This sequence, rhythm four, consists of coarse conglomeratic channel fill at its base followed by gray channel sands and siltstones, red flood-basin sands and siltstones, red overbank mudrocks, and gray, olive, and black mudrocks of the marginal swamps and marshes. The sequence is set as a carpet of braided stream deposits on the subjacent strata (Buttner, 1967).

In this model, data on 10 sedimentological parameters are presented graphically for 15 stations located in the North Point area of Greene County, New York. The dashed lines show the outcrop limit with each station located at the apex of a pebble composition triangle. At each station, pebbles (>32 mm) were examined, measured, and classified until at least 1,000 pebbles of various sedimentary rock types were counted (total sample sizes ranged from 1,500 to 3,800 clasts). These 1,000 pebbles were put into fluvial (F), shelf (S), or turbidite (T) classes using thin sections as required. At every station the composition triangle shows the proportion of each of these provenance domains represented at that site. Bar graph 1

ANALYTIC MODELS

The algorithm on which the analytic model is based determines the extended utility of the model. This algorithm can be a single equation, groups of equations, statistical procedures, numerical techniques, logical expressions, interactive machines, a sequence of models, or some combination of these. In some studies, as with a sufficient descriptive model, the development of an analytic model completes the analysis. In other investigations a properly designed and executed analytic model can be generalized and extended into the development of a simulation model. As shown in Figure 7, two main differences between analytic and simulation models are that the former considers a single picture of a system stated in terms of tractable (solvable) analytic procedures, and the latter attempts to describe a sequence of "snapshots" using both tractable and probabilistic arguments. With this view we picture the simulation model as a sequence of analytic models; at the moment of solution (when a probabilistic input is given), the simulation model is tractable. Some workers extend the term "simulation" to include any kind of tractable solution (for example, the solution of $y = mx + b$ simulates a tractable condition). However, terminology is not as important as ontogeny, as simulation models are usually developed from tractable analytic structures (Buttner, 1966b).

Linear programming, a technique useful in the development of analytic models (Buttner, 1967), should perhaps be called linear scheduling; although computers and computer programs are used in its execution, it is not computer programming. Gass (1969) is a prime source of information on this technique. The linear programming model consists of three parts: (1) a statement of the objective to be optimized; (2) a statement of the conditions of optimization; and (3) a statement of the relations among those conditions. For instance, consider the fluvial blending model in Figure 12. In this model there are m sources of sand (actually a set of grain-size frequency distributions) and n lithosomes. The total amount of sand available from the ith source is given by a_i, and the total requirement of sand

---

shows the ratio of these sedimentary pebbles (dark bar) to pebbles of all other domains collected at the site. The mean of the long axes of the 10 largest pebbles collected was compared to that of the largest collected at that site to establish the maximum pebble size (black circle at triangle apex). The attached arrow shows the summary of all directional features (clast orientations, bed-form geometry, channel shape, local trends) for that station. The second bar graph gives the normalized thickness (stippled) of the rhythm for that station. Bar graph 3 illustrates the ratio of the total thickness of the conglomerate + sandstone (diagonal bars) to the total thickness of siltstone + mudrocks in the rhythm at each station. The ratio of red matrix rocks (horizontal bars) to the thickness of gray matrix rocks is shown in bar graph 4. At each station the thickest of the 10 thickest fore-set beds was compared with their mean to get a station maximum. This value was normalized and is shown as the stippled area in bar graph 5. Bar graph 6 shows the proportion of each of four bed-form families: flatbeds (upper triangle), ripples (next triangle), dunes (next triangle), and antidunelike structures (lowest triangle). The normalized thickness of the rhythm at each station is given in bar graph 7 (vertical lines). The collection of these data followed appropriate experimental design procedures (Buttner, 1965, 1968).

at each lithosome is given by b_j. X_{ij} is the amount of sand from source i that reaches lithosome j; C_{ij} is an expression that describes the textural relations (amounts of crossbedding, mean grain size, lithologic ratios, bedding thickness, directional properties, and so forth) required if one module of sand from source i is used to build lithosome j.

Linear programming has been used to design and evaluate a number of three-dimensional synthetic models. Models have been built of fluvial networks, deltas, sampling plans, faunal stratifications, and of many physical-stratigraphic domains. To illustrate its use in building a model of fluvial basin deposits we can use a modified standard problem from the literature for finding the maximum optimal solution. The experimental setting for this much-simplified problem is as follows:

1. A section contains 72 units of mudrocks, 180 units of sandstone, and 90 units of conglomerate; a unit is some appropriate linear or textural index.

2. We recognize both mainstream and flood-basin rhythmic sequences in the section composed of these lithologies.

3. Mainstream sequences are composed of 3 units of conglomerate, 5 units of sandstone, and 1 unit of mudrocks.

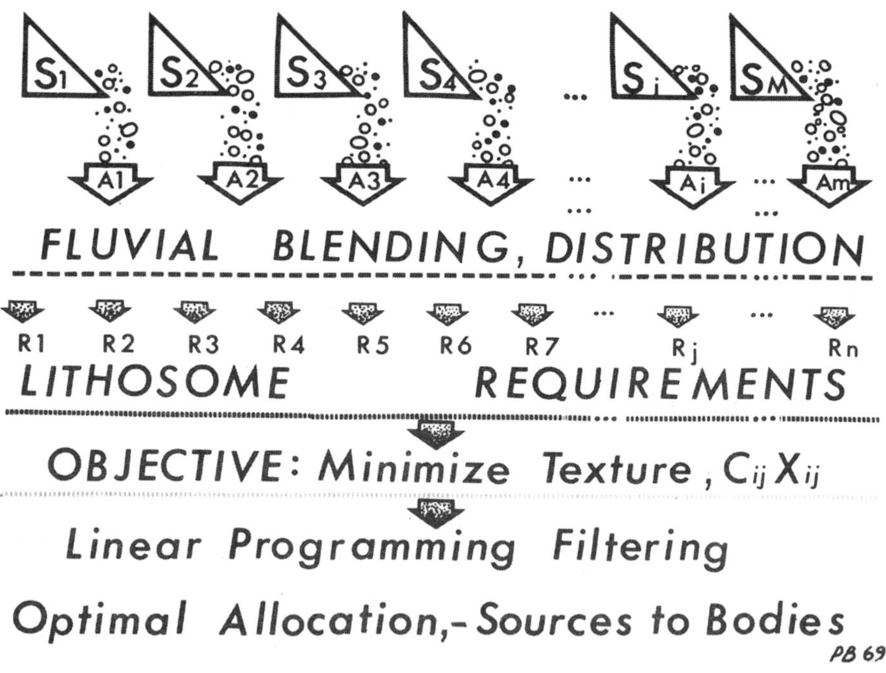

Figure 12. Linear programming model. Each source, S, can provide the amount, A, of "texture builder" (grain-size frequency distributions, shape populations, mineralogical propensities, and so forth) to be used to build a particular lithosome. The development of each lithosome follows a set of source, accumulation, distribution, and preservation—disciplines according to its requirements, R, and with the structural and functional objectives specified by the constraints, CX (Buttner, 1969a).

4. Flood-basin sequences are composed of 1 unit of conglomerate, 4 units of sandstone, and 2 units of mudrocks.

5. We wish to find the maximum number of both sequences that we can expect on the basis of optimal allocations of rock units.

The problem statement and solution details are shown in Figure 13. The optimal allocation of response events is to use 12 mainstream sequences and 30 flood-basin sequences; the leftover conglomerate might be considered lag deposits. Notice that this method only permits the model builder to synthesize structures. In this context important sedimentologic and stratigraphic questions are not considered in the solution; increased specification of the allocation setting is required to involve these aspects. While very much simplified, this example shows how linear programming can be used by a stratigrapher as a powerful tool for the development of an analytic model.

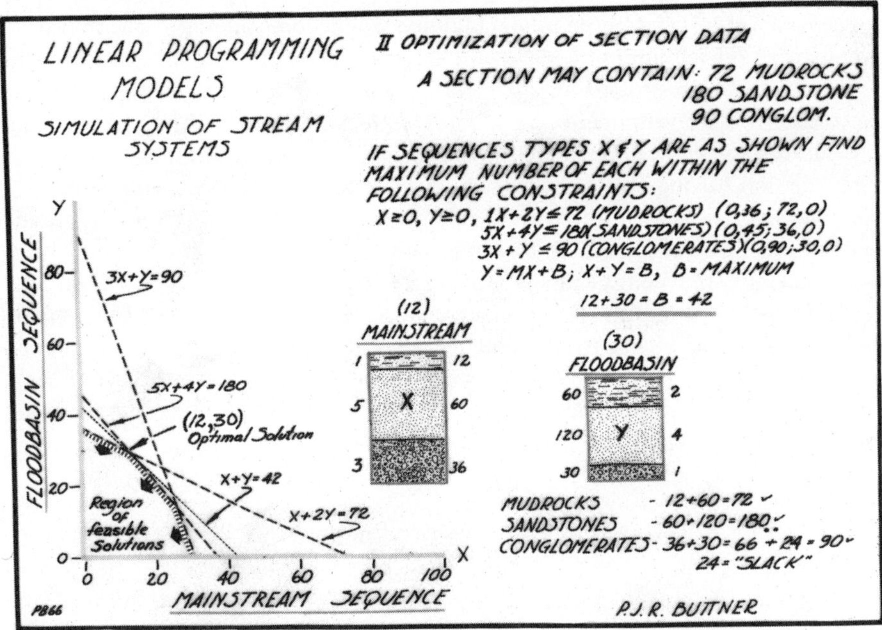

Figure 13. Lithosome modeling. A section contains three lithologic types (conglomerate, sandstone, and mudrocks), organized into two fluvial sequences (mainstream and flood basin). Mainstream sequences contain the lithologies in the proportions 3, 5, and 1, and flood-basin sequences in the proportions 1, 4, and 2 (conglomerate, sandstone, and mudrocks, respectively). The region of feasible solution is set off by the intersection of equations. The point where the number of mainstream sequences plus the number of flood-basin sequences (X + Y) is maximum occurs at (12, 30), the optimal solution. This means the maximum number of sequences we can extract from this section is 12 mainstream and 30 flood basin. Numerical calculations based on this optimum are shown. A three-dimensional lithosome may be assembled by simultaneous solution of sections in the same stratigraphic unit. Experimental setting and problem specification have been greatly simplified. (Example *from* Buttner, 1967.)

SUMMARY

The subject of this work is the building of symbolic models. I have attempted to define the workshop in which the operation of model building proceeds best: systems analysis. To illustrate this argument we have examined a variety of symbolic models. Most of these models were constructed within the framework of some operational research techniques such as linear programming. I could also have used illustrative models that are based on statistical techniques, numerical analysis, or other methodologies. The operational research format was chosen because it is less commonly encountered in the training program of most geologists. Additionally, there are other models, such as some of the simulation types that use a suite of techniques producing a model structure considerably more complex than those presented here. Some simulation models use gaming as part of their structure and permit an instructor and his basin model to interact with his class of students via a remote computer. I expect that this type of interactive computer-assisted instruction on model building will become an important part of the geometrics curriculum in the near future.

Perhaps the most demanding, exciting, and rewarding task of the systems investigator is his development of a simulation model (Fig. 14). The builder of such models cannot know enough about his subject. Because of this he derives a unique appreciation for the subtle nature of his system and develops a keen rapport with the primary aspects of its structure and function. His contribution is especially valuable to his science because he can provide a reservoir of expertise previously not available. In recent years the model-building activities of J. C. Griffiths, G. M. Friedman, L. I. Briggs, W. C. Krumbein, and L. L. Sloss have resulted in a significantly greater sensitivity to the fundamental properties of stratigraphic systems. If more workers were to develop an appropriate symbolic model as part of their research, our ability to define the *structure* and analyze the *function* of many geologic systems would be significantly enhanced.

ACKNOWLEDGMENTS

This work was taken in part from a thesis submitted in partial fulfillment of the requirements for the degree of Doctor of Philosophy in the Department of Geology at Rensselaer Polytechnic Institute, Troy, New York.

Field, laboratory, and other expenses were defrayed in part by the Society of the Sigma Xi, the American Association of Petroleum Geologists, the National Science Foundation (fellowships and Grant GA 300), the University of Rochester, Rensselaer Polytechnic Institute, and Union College.

During the last ten years I have received various forms of assistance in my simulation experiments from E. G. Blum, S. Buttner, G. Buttner, T. A. Keenan, L. Palmer, R. Sack, L. Smith, R. G. Sutton, V. H. Swoyer, and S. Weaver. I am pleased to acknowledge the inspiration for, and the review of, many of those experiments by Natalia Buttner, L. I. Briggs, G. M. Friedman, G. V. Middleton, R. A. Park, M. P. Wolff, and C. Wachmann.

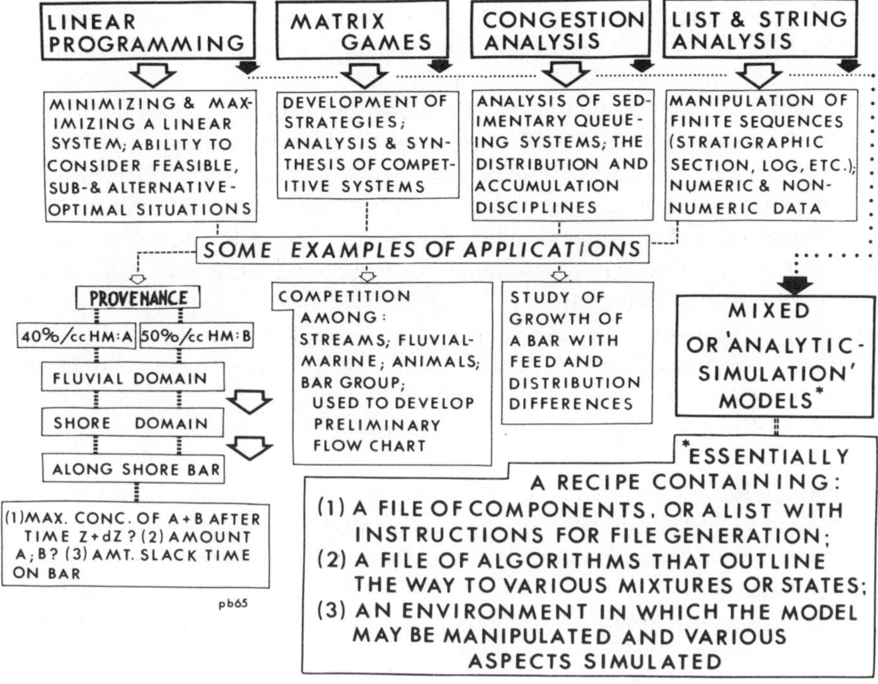

Figure 14. Operational research techniques in geology. Operational research methods such as linear programming, matrix gaming, congestion analysis, and list processing can be used in the formulation, design, and evaluation of models of geologic systems. These kinds of techniques are used primarily to develop models which define the *structure* of a system; these models are then usually assembled within the framework of simulation model to analyze the *function* of that system. The simulation model consists of three parts: an organized data resource, a collection of simulation techniques, and a simulation environment (Buttner, 1966b).

Portions of this manuscript have been reviewed by colleagues at North American Rockwell, New York Department of Transportation, Rensselaer Polytechnic Institute, Union College, the General Electric Company, and Mechanical Technology Incorporated.

REFERENCES CITED

Buttner, P.J.R., 1965, Response model design for a rhythmic delta platform domain, Devonian Catskill Complex of New York [abs.]: Am. Assoc. Petroleum Geologists Bull., v. 49, p. 336.
—— 1966a, Introduction to geologic models, *in* Briggs, L. I., and Pollack, H. N., Computer techniques for the petroleum geologists: Ann Arbor, Univ. Michigan Eng. Conf., Proc. and Notes.

Buttner, P.J.R., 1966b, Simulation models in stratigraphy and sedimentation, *in*
    Briggs, L. I., and Pollack, H. N., Computer techniques for the petroleum geol-
    ogists: Ann Arbor, Univ. Michigan Eng. Conf., Proc. and Notes.
—— 1967, An introduction to linear programming models in geology, *in* Briggs,
    L. I., Data processing and computer modeling for geologists: Ann Arbor,
    Univ. Michigan Eng. Conf., Proc. and Notes.
—— 1968, Proximal continental rhythmic sequences in the Genesee Group
    (Lower Upper Devonian) of southeastern New York, *in* Klein, G. deVries,
    ed., Late Paleozoic and Mesozoic continental sedimentation, northeastern
    North America: Geol. Soc. America Spec. Paper 106, p. 109–126.
—— 1969a, Molasse, managers, and model geology: analysis, control and design
    with the aid of linear programming: Geol. Soc. America, Abs. with Programs
    for 1969, Pt. 7 (Ann. Mtg.), p. 27.
—— 1969b, Textural evaluation of aggregate materials in Portland cement con-
    crete: Highway Research Board, Committee MCB2, Performance of Con-
    crete–Chemical Aspects.
—— 1972, The limnology of Glass Lake, Rensselaer County, New York (in
    press).
Buttner, P.J.R., and Wachmann, C., 1972, Ecosensitive municipal water strategies:
    A model study of the Norman's Kill drainage basin, eastern New York
    State: Geol. Soc. America, Abs. with Programs (Northeastern Sec.), v. 4,
    no. 1, p. 6.
Duncan, A. J., 1965, Quality control and industrial statistics: Illinois, Richard D.
    Irwin, 992 p.
Gass, S. I., 1969, Linear programming: New York, McGraw-Hill Book Co., 358 p.
Griffiths, J. C., 1967, Scientific method in analysis of sediments: New York, Mc-
    Graw-Hill Book Co., 508 p.
Harbaugh, J. W., and Bonham-Carter, Graeme, 1970, Computer simulation in geo-
    logy: New York, John Wiley & Sons, 575 p.
Krumbein, W. C., and Graybill, F. A., 1965, An introduction to statistical models
    in geology: New York, McGraw-Hill Book Co., 475 p.

MANUSCRIPT MODIFIED FROM TALK GIVEN AT QUANTITATIVE GEOLOGY SYM-
POSIUM AT THE GEOLOGICAL SOCIETY OF AMERICA ANNUAL MEETING IN
ATLANTIC CITY, NOVEMBER 10, 1969

Cybernetics-Geomathematics Interaction

John C. Griffiths
Department of Geochemistry and Mineralogy
College of Earth and Mineral Sciences
The Pennsylvania State University
University Park, Pennsylvania 16802

ABSTRACT

Adoption of new tools to solve old problems evolves in several stages. First, there are sporadic forays, such as Udden's initial quantification of grain-size analysis (1898), or Chamberlin's geological epistemology (1897), or Sorby's sedimentary petrography (1880). This is followed by a systematic attack in which some pioneer defines a new analytical mode and many follow his lead; Krumbein's attack on sedimentary petrography in the 1930s exemplifies this stage and many petrographers adopted his approach. A third stage leads to the establishment of a new field of specialization, such as sedimentology, x-ray crystal chemistry, geochemistry, and geomathematics. Even at this stage most effort is concentrated on solving old problems with new techniques. However, the advantages of new tools are lost in this approach; the elegance and main advantage of mathematics lie in its rigor, and the rigor is lost as the analytical tools are bent to fit ill-defined problems. In the next stage, problems are reformulated, and this necessitates new paradigms and at this stage there is real breakmulated, and this necessitates new paradigms and at this stage there is real breakthrough.

Geomathematics (quantitative geology) appears to be at this threshold; future geoscience questions embrace people-problems and de-emphasize the inanimate. These exceedingly complex probabilistic problems of man-machine symbiotic systems require such tools as operations research, cybernetics, and systems analysis. Problems are now redefined in a metalanguage or they become unresolvably Goedelian. Thus, adoption of new tools, by feedback, necessitates redefinition of old problems, adoption of new paradigms, leading to self-renewal of the (geoscience) field of operations.

INTRODUCTION

The scientific method is an evolving procedure, progressing in steps as new software and hardware are devised (Griffiths, 1968, Fig. 1). Progress arises largely from pressures which demand renewed appraisals of the scientific enterprise (*see*, for example, Kuhn, 1962). Quantification, first recommended and used by illustrious geoscience predecessors (Lyell, 1833; Mackie, 1897; Sorby, 1880), is becoming a popular practice in geoscience problem-solving; it was not until the statistical method, introduced by Krumbein in the mid-1930s, became popular in the 1950s, that quantification caught on (with all due deference to the geophysicists who were quantitative by instinct and training from the beginning, but, in a mathematical mode described as deterministic rather than stochastic and that relies on continuity rather than quantum jumping). Nowadays, ready access to computers leads to electronic data processing and accelerates the applications of statistics by reducing computational drudgery and automating statistical analysis.

Once this breakthrough occurred, enthusiasm led to spillover and the adoption of many relatively new tools of analysis, such as operations research, information theory, cybernetics, and systems analysis. Much of this activity is first tried and encouraged by the mineral industries, an area of applied geoscience that relies on competitive enterprise for its continued existence.

As might be expected, the main pioneering efforts with these new tools are practiced by an enthusiastic minority and do not reflect the bulk of the research in progress. When the current geoscience literature is examined, we find a strong tendency to replace long, involved descriptions by numerals, adoption of a symbolism which cannot help but improve investigation if only by reducing the volume of output. It is, however, noteworthy that the manipulation of these numerals is not often indulged in so that they are not really looked upon as numbers, often quite rightly so; analysis is therefore not nearly so common (Griffiths, 1970).

In most cases new forms of analysis are applied in an attempt to solve some existing geoscience problem, and quite often, because the analytical tool is not tailored to fit the problem, it is usually bent a little to fit this special case. Many of the analytical tools are mathematical in form, and one of the advantages of mathematics is its rigor; unfortunately, bending the tool usually relaxes the rigor and much of the advantage sought in the use of mathematics is lost. An alternative approach would be to find a subject-matter problem that matches the requirements of the tool or even to redefine the problem to match the tool; to practicing scientists in most subject-matter areas, this appears to be rank heresy, although there are other views, for example, "science might almost be defined as the process of substituting unimportant questions which can be answered for important questions which cannot" (Boulding, 1961, p. 164).

Perhaps the time has come to examine our geoscience problems with reformulation in mind, not only because this may tend to rejuvenate the science by redefinition of its paradigms (Kuhn, 1962), but also because the relevant problems are no longer solely concerned with inanimate rocks or lowly forms of past life! Once the problems encompass present-day (relevant?) man-machine systems, judgments of value enter the field, and the unsuspecting "pure" scientist is sooner or later trapped into a social or political context for which his training has made him singularly ill-fitted. Under these conditions, most of the science we learned before 1960 appears somewhat naïve, and the scientific method of problem solving is now compelled to proceed beyond statistical testing to embrace decision theory, game theory, information processing, and the like. In an operations research context, there is usually an objective function to be maximized and this includes a judgment of value. This procedure need not necessarily depend upon or be translated into dollar value; indeed, it may be so designed as to include solely inanimate aspects, as illustrated by Buttner (this volume). What the modern aspects of the scientific method have accomplished in embracing these new tools is to make explicit what is practiced (and not solely what is preached!).

SOME ASPECTS OF PROBLEM-SOLVING USING THE NEW MODES

The restless drive to adopt ever more sophisticated forms of analysis also gives rise to questions that until now have appeared to be unimportant, already resolved, or too exotic to be of interest. For example, a geologist is an expert in studying the past, but then so is an archaeologist and a historian; how is it that we can so confidently describe the events of millions of years ago although we have difficulty with archaeological events and very considerable difficulty with historical events that occurred less than 2,000 years ago?

Even more striking: why is it so simple to predict the long-distant past by means of the principle of uniformitarianism—the present is the key to the past—and almost impossible to predict the future? Would anyone attempt to apply uniformitarianism to future events and, if not, why not, and what makes it then so powerful in the past?

As mentioned elsewhere, if past events are judged to be parts of sequences that may be described as regular ergodic Markov processes, then indeed the present is not much of a key to the past (Griffiths, 1970). Indeed, cyclic processes that are now modelled on Markov processes generally possess very limited memories (Carr, 1966; Krumbein, 1968; Potter and Blakely, 1968); if this peculiar feature is combined with the property systems theorists call "equifinality," we are indeed led into something of a paradox (von Bertalanffy, 1968). Equifinality essentially implies that a system (for example, an organism) may arrive at a final stage in evolution from many different starting points or by many differing process steps, or both, and if we examine the system at this stage, it is not possible to determine which starting point or which path the system followed. Examples in a geoscience context are quite common; for example, a detrital quartzite, as defined by Krynine (1948), may be an excessively refined arkose, or low-rank or high-rank graywacke, and it may not be possible to decide the evolutionary path (Griffiths and Ondrick, 1969); many carbonates, which have suffered diagenetic alteration, contain no remaining evidence of original structure and, therefore, of process. Many metamorphic rocks must be in analogous stages, and their past history is unlikely to yield to present analytical procedures or uniformitarian philosophy.

It seems likely that examples may abound in paleontology (homeomorphs?) and are strictly analogous to the example of the starfish described by von Bertalanffy (1968, p. 40).

Examples in the quantitative mode are also likely to be frequent, as when a stratigraphic sequence of rocks is described in the field, represented as a Markov process, found to be ergodic regular and then the final matrix is a constant vector (Griffiths, 1966). It is clear that no matter what sequence is observed, the end point is equifinal and appears to have no unique memory of the exact steps involved in its achievement.

The Goedel Incompleteness Theorem represents another obstacle to analysis in the conventional mode; first published in 1931, it has taken some time to percolate downward so that it may be commonly comprehended (Nagel and Newman, 1958). The relevant implication, in the present context, concerns the feature that within any consistent system questions arise that are undecidable within the system although their resolution appears feasible from outside the system. Practical application of this issue is exemplified by Beer (1964) in the context of management, and he indicates that the problem may be circumvented, if not resolved, by the "principle of completion from without" (Beer, 1966, p. 288). This may be loosely translated to mean that within any scientific system, such as geoscience, with its axioms and principles of operation, questions may arise that are unresolvable within this system, but which may be seen to be resolvable from outside the system. Essentially, for example, the development of geophysics broadens the geoscience system so that unresolvable problems that may be illuminated by physics are now within the scope of geoscience; to comprehend geophysics, it is necessary for the geologist to learn a new language (or jargon, if preferred), and this is essentially the way around the incompleteness of the initial system. A metalanguage should be devised that will permit the problem solver to reformulate the question in such a way that it may be resolved in terms of the metalanguage. In a sense, the growth of geomathematics is equivalent to the development of a new metalanguage for the geosciences, and many hitherto unresolvable questions may be resolved in terms of this metalanguage of geoscience.

The problem of continental drift is an example in which, when Wegener (1924) proposed it as a process to explain the distribution of land and water masses through time, most of the evidence for drift was established, but the proposition of drifting was unresolvable in the light of contemporary earth mechanics and earth structure of the time. Now, with the spreading of ocean floors as a mechanism for drift, the hypothesis has taken on new life and may become a central principle in explaining the origin of the earth's surface features.

It seems likely that many other problems that appear unresolvable within the present system of geoscience may yield to solution when a suitable systems metalanguage is devised that will suffice to complete the system from without (Beer, 1964).

Many questions that appear to be of Goedelian character loom on the horizon of present-day geoscience, particularly problems that concern the public interest, such as pollution of air, sea, and land and the conservation and development of natural resources, marine and nonmarine. In these cases, choice of priorities will require some very difficult decisions about alternative values and will clearly demand a new metalanguage for their solution—a language that will be logically rigorous and suitably weighted with human opinion based on future values. All these features have been present in past questions but their explicit recognition and consideration are relatively new.

This may give rise to still another obstacle described by Ashby (1964) that, in its present form, appears to put an upper limit on achievement within the present system of science, let alone geoscience. Ashby considers the magnitude of alternatives that arise as we contemplate the interactions within a real-world system; he deduces the following "law": *everything material stops at* 10^{100}. He goes on to show that using (as a simple example) a square block of lamps measuring 20×20 (that is, 400 lamps) in which each lamp may be on or off, there are 2^{400} alternative pictures; this is about equivalent to 10^{120}, already exceeding very comfortably the limit of 10^{100}. Second, suppose the pictures are of two kinds, A and *not-A*, then the number of properties from which any particular set is picked out is $2^{(10^{120})}$ or approximately $10^{10^{120}}$. He then proceeds to show that such a number is, at present, well beyond the physically achievable; he further demonstrates that the number of properties definable on the block of 20×20 lamps, $10^{10^{120}}$, is a number "not appreciably affected when divided by a number itself so large it cannot be written in our universe" (Ashby, 1964, p. 167). Now, how many problems in a natural science may be considered to possess say 20×20 variables of the go-no-go type? Most real-world geoscience problems would often comfortably exceed this example, and hence the interactions increase in complexity, orders of magnitude beyond what we usually think of in typically scientific problems, particularly in those we are inclined to think of as solved!

In an earlier article, Ashby (1956) introduced the idea of an "intelligence amplifier," a complex man-machine system that, if appropriately devised, would serve to amplify intelligence in an analogous manner to the muscle amplification by machines with which we are now so familiar; this device may serve to develop potential mental capacity so that problems of the complexity above described may be within our comprehension. It is interesting to consider that such intelligence amplifiers, complex man-machine systems for information processing, are potentially in existence, namely in geological surveys, bureaus of mines, educational institutions, and the like, although these may not now function quite as effectively as Ashby's concept implies or as the solution to these exceedingly complex problems requires.

GEOMATHEMATICS-CYBERNETICS INTERACTION

The growth of geomathematics likely to be typical of the 1970s (Griffiths, 1970), together with the change in environment of problem-solving in the geosciences that is also presaged for the next decade, will require considerable change in our approach to geoscience problem-solving. That is, the scientific method now in general use will be modified by extension to include interaction of man-machine cybernetic systems. The problems themselves will be different in terms of the complexity of the interactions and will be of the type associated with "exceedingly complex probabilistic systems" (Beer, 1964, p. 18; Griffiths, 1970, Fig. 1). Some idea of what this portends is exhibited in Figure 1; any subject-matter area tends to be a self-consistent logical system with its own paradigms (Kuhn, 1962). By use of these paradigms, we formulate the subject-matter problems and then proceed by the conventional scientific method to solve the problems (for examples in sedimentary petrography, *see* Griffiths, 1967). These steps are included within the box labelled geostatistics in Figure 1. We are inclined to stop at the acceptance or rejection of the (Null) hypothesis, and all feedbacks occur within this subsystem. These steps in analysis may suffice to comprehend problems associated with simple deterministic, complex deterministic, and simple probabilistic systems (Beer, 1964, p. 12ff; Griffiths, 1970, Fig. 1). However, when, because of environmental pressure, we are compelled to consider problems concerned with complex probabilistic or extremely complex probabilistic systems, this type of analysis is likely

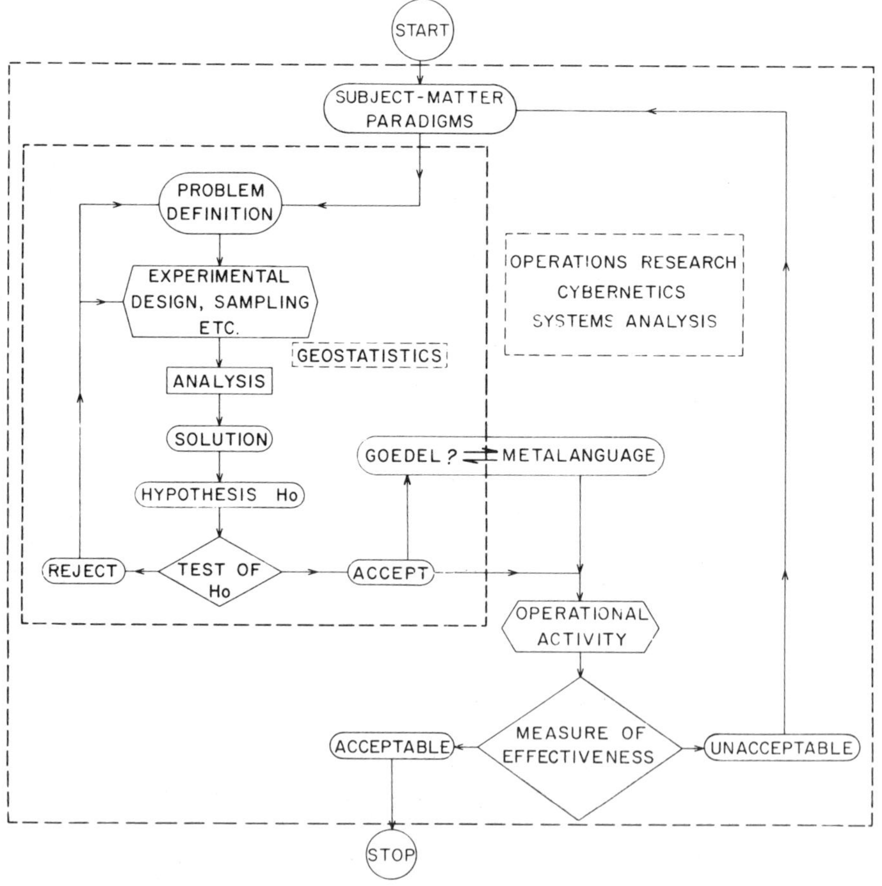

Figure 1. **Flow diagram illustrating the geomathematics-cybernetics interaction.**

to prove inadequate; it is then necessary to extend the method, to devise an operational procedure, and the outcome of this procedure is judged against a measure of effectiveness that is either acceptable or unacceptable to the people using our results. If unacceptable, the feedback loop leads to questioning of the paradigms and a complete change in the underlying foundations of the subject-matter area; examples are described by Kuhn (1962) in the "pure" science milieu. Examples in geoscience arise in exploration for natural resources and will be much more likely to arise with future emphasis on people problems, such as in earthquake prediction.

Suppose that at the stage of hypothesis testing the alternative is not only accept-reject but also raises the question of Goedelian Incompleteness (*see* Fig. 1), then it may be necessary to construct a metalanguage before any solution is achievable, and this, in turn, leads to an operational activity and so on. It is conceivable that many questions now considered adequately solved will, when put to operational testing, prove unacceptable and indeed will turn out to be Goedelian questions. The metalanguage will interact with the subject-matter paradigms, and as suggested in the above account, we may have to modify and even abandon some of our beloved traditional principles.

It seems, then, that geoscience, through its interactions with the new tools of the extended scientific method, such as the geomathematics $\rightleftharpoons$ cybernetics interaction, will enter its new role in attacking relevant problems of society. This maturation likely will lead to questioning of its traditional paradigms and to the adoption of new principles that, in turn, will rejuvenate geoscience. In terms of systems science, geoscience will prove viable by exhibiting its characteristics as a self-organizing system interacting with its changing environment.

REFERENCES CITED

Ashby, W. R., 1956, Design for an intelligence amplifier: Automata Studies, Princeton Univ. Press.

—— 1964, Introductory remarks at panel discussion, *in* Mesarovic, M. D., ed., Views on general system theory: Proc. 2d., Systems Symposium, Case Inst. Technology, New York, John Wiley & Sons, 178 p.

Beer, S., 1964, Cybernetics and management: Science Editions, New York, John Wiley & Sons, 214 p.

—— 1966, Decision and control: the meaning of operational research and management cybernetics: New York, John Wiley & Sons, 556 p.

Boulding, K. E., 1961, The image: Univ. Michigan Press, Ann Arbor Paperbacks, 175 p.

Carr, D. D., 1966, Stratigraphic sections, bedding sequences, and random processes: Science, v. 154, p. 1162–1164.

Chamberlin, T. C., 1897, The method of multiple working hypotheses: Jour. Geology, v. 5, p. 837–848 (reprinted in Jour. Geology, v. 39, p. 155–165, 1931).

Griffiths, J. C., 1966, Future trends in geomathematics: Mineral Industries, v. 35, p. 1–8.

Griffiths, J.C., 1967, Scientific method in analysis of sediments: New York, McGraw-Hill Book Co., 508 p.

—— 1968, A review of operations research in the mineral industries: Western Miner, v. 41, p. 22–26.

—— 1970, Current trends in geomathematics: Earth-Science Reviews, v. 6, p. 121–140.

Griffiths, J. C., and Ondrick, C. W., 1969, Modelling the petrology of detrital sediments, *in* Merriam, D. F., ed., Computer applications in the earth sciences: New York, Plenum Press, 281 p.

Krumbein, W. C., 1968, Statistical models in sedimentology: Sedimentology, v. 10, p. 7–23.

Krynine, P. D., 1948, The megascopic and field classification of sedimentary rocks: Jour. Geology, v. 56, p. 130–165.

Kuhn, T. S., 1962, The structure of scientific revolutions: Univ. Chicago Press, 172 p.

Lyell, C., 1833, Principles of geology (2d ed.): London, J. Murray, v. 3.

Mackie, W., 1897, On the laws that govern the rounding of particles of sand: Edinburgh, Geol. Soc. Trans., v. 7, p. 298–311.

Nagel, E., and Newman, J. R., 1958, Gödels proof: New York Univ. Press, 118 p.

Potter, P. E., and Blakely, R. F., 1968, Random processes and lithologic transitions: Jour. Geology, v. 76, p. 154–170.

Sorby, H. C., 1880, The structure and origin of non-calcareous stratified rocks: Geol. Soc. London Quart. Jour., v. 36, p. 46.

Udden, J. A., 1898, The mechanical composition of wind deposits: Augustana Library Pub. 1.

von Bertalanffy, L., 1968, General system theory: New York, G. Braziller, 289 p.

Wegener, A., 1924, The origin of continents and oceans: translated by J.G.A. Skerl, London, Methuen and Co., Ltd., 212 p.

MANUSCRIPT MODIFIED FROM TALK GIVEN AT QUANTITATIVE GEOLOGY SYMPOSIUM AT THE GEOLOGICAL SOCIETY OF AMERICA ANNUAL MEETING IN ATLANTIC CITY, NOVEMBER 10, 1969

Index